D0627019

THE MEDITATIONS

OF

DESCARTES

Seventh Edition

Translated From The Latin & Collated With The French
by John Veitch

Watchmaker Publishing
1880

ISBN 978-1-60386-391-9

DIGITALIZED BY
WATCHMAKER PUBLISHING

CONTENTS.

I.—THE MEDITATIONS.

TO

THE VERY SAGE AND ILLUSTRIOUS

THE

DEAN AND DOCTORS OF THE SACRED FACULTY OF THEOLOGY OF PARIS.

GENTLEMEN,

The motive which impels me to present this Treatise to you is so reasonable, and, when you shall learn its design, I am confident that you also will consider that there is ground so valid for your taking it under your protection, that I can in no way better recommend it to you than by briefly stating the end which I proposed to myself in it. I have always been of opinion that the two questions respecting God and the Soul were the chief of those that ought to be determined by help of Philosophy rather than of Theology; for although to us, the faithful, it be sufficient to hold as matters of faith, that the human soul does not perish with the body, and that God exists, it yet assuredly seems impossible ever to persuade infidels of the reality of any religion, or almost even any moral virtue, unless, first of all, those two things be proved to them by natural reason. And since in this life there are frequently greater rewards held out to vice than to virtue, few would prefer the right to the useful, if they were restrained neither by the fear of God nor the expectation of another life; and although it is quite true that the existence of God is to be believed since it is taught in the sacred Scriptures, and that, on the other hand, the sacred Scriptures are to be believed because they come from God (for since faith is a gift of God, the same Being who bestows grace to enable us to believe other things, can likewise impart of it to enable us to believe his own existence), nevertheless, this cannot be submitted to infidels, who would consider that the reasoning proceeded in a circle. And, indeed, I have observed that you, with all the other theologians, not only affirmed the sufficiency of natural reason for the proof of the existence of God, but also, that it may be inferred from sacred Scripture, that the knowledge of God is much clearer than of many created things, and that it is really so easy of acquisition as to leave those who do not possess it blame-worthy. This is manifest from these words of the Book of Wisdom, chap. xiii., where it is said, *Howbeit they are not to be excused; for if their understanding was so great that they could discern the world and the creatures, why did they not rather find out the Lord thereof?* And in Romans, chap. i., it is said that they are *without excuse*; and again, in the same place, by these words,—*That which may be known of God is manifest in them*—we seem to be admonished that all which can be known of God may be made manifest by reasons obtained from no other source than the inspection of our own minds. I have, therefore, thought that it would not be unbecoming in me to inquire how and by what way, without going out of ourselves, God may be more easily and certainly known than the things of the world.

And as regards the Soul, although many have judged that its nature could not be easily discovered, and some have even ventured to say that human reason led to the conclusion that it perished with the body, and that the contrary opinion could be held through faith alone; nevertheless, since the Lateran Council, held under Leo X. (in session viii.), condemns these, and expressly enjoins Christian philosophers to refute their arguments, and establish the truth according to their ability, I have ventured to attempt it in this work. Moreover, I am aware that most of the irreligious deny the existence of God, and the distinctness of the human soul from the body, for no other reason than because these points, as they allege, have never as yet been demonstrated. Now, although I am by no means of their opinion, but, on the contrary, hold that almost all the proofs which have been adduced on these questions by great men, possess, when rightly understood, the force of demonstrations, and that it is next to impossible to discover new, yet there is, I apprehend, no more useful service to be performed in Philosophy, than if some one were, once for all, carefully to seek out the best of these reasons, and expound them so accurately and clearly that, for the future, it might be manifest to all that they are real demonstrations. And finally, since many persons were greatly desirous of this, who knew that I had cultivated a certain Method of resolving all kinds of difficulties in the sciences, which is not indeed new (there being nothing older than truth), but of which they were aware I had made successful use in other instances, I judged it to be my duty to make trial of it also on the present matter.

Now the sum of what I have been able to accomplish on the subject is contained in this treatise. Not that I here essayed to collect all the diverse reasons which might be adduced as proofs on this subject, for this does not seem to be necessary, unless on matters where no one proof of adequate certainty is to be had; but I treated the first and chief alone in such a manner that I should venture now to propose them as demonstrations of the highest certainty and evidence. And I will also add that they are such as to lead me to think that there is no way open to the mind of man by which proofs superior to them can ever be discovered; for the importance of the subject, and the glory of God, to which all this relates, constrain me to speak here somewhat more freely of myself than I have been accustomed to do. Nevertheless, whatever certitude and evidence I may find in these demonstrations, I cannot therefore persuade myself that they are level to the comprehension of all. But just as in geometry there are many of the demonstrations of Archimedes, Apollonius, Pappus, and others, which, though received by all as evident even and certain (because indeed they manifestly contain nothing which, considered by itself, it is not very easy to understand, and no consequents that are inaccurately related to their antecedents), are nevertheless understood by a very limited number, because they are somewhat long, and demand the whole attention of the reader: so in the same way, although I consider the demonstrations of which I here make use, to be equal or even superior to the geometrical in certitude and evidence, I am afraid, nevertheless, that they will not be adequately understood by many, as well because they also are somewhat long and involved, as chiefly because they require the mind to be entirely free from prejudice, and able with ease to detach itself from the commerce of the senses. And, to speak the truth, the ability for metaphysical studies is less general than for those of geometry. And, besides, there is still this difference that, as in geometry, all are persuaded that nothing is usually advanced of which there is not a certain demonstration, those but partially versed in it err more frequently in assenting to what is false, from a desire of seeming to understand it, than in denying what is true. In philosophy, on the other hand, where it is believed that all is doubtful, few sincerely give themselves to the search after truth, and by far the greater number seek the reputation of bold thinkers by audaciously impugning such truths as are of the greatest moment.

Hence it is that, whatever force my reasonings may possess, yet because they belong to philosophy, I do not expect they will have much effect on the minds of men, unless you extend to them your patronage and approval. But since your Faculty is held in so great esteem by all, and since the name of SORBONNE is of such authority, that not only in matters of faith, but even also in what regards human philosophy, has the judgment of no other society, after the Sacred Councils, received so great deference, it being the universal conviction that it is impossible elsewhere to find greater perspicacity and solidity, or greater wisdom and integrity in giving judgment, I doubt not,—if you but condescend to pay so much regard to this Treatise as to be willing, in the first place, to correct it (for, mindful not only of my humanity, but chiefly also of my ignorance, I do not affirm that it is free from errors); in the second place, to supply what is wanting in it, to perfect what is incomplete, and to give more ample illustration where it is demanded, or at least to indicate these defects to myself that I may endeavour to remedy them; and, finally, when the reasonings contained in it, by which the existence of God and the distinction of the human soul from the body are established, shall have been brought to such degree of perspicuity as to be esteemed exact demonstrations, of which I am assured they admit, if you condescend to accord them the authority of your approbation, and render a public testimony of their truth and certainty,—I doubt not, I say, but that henceforward all the errors which have ever been entertained on these questions will very soon be effaced from the minds of men. For truth itself will readily lead the remainder of the ingenious and the learned to subscribe to your judgment; and your authority will cause the atheists, who are in general sciolists rather than ingenious or learned, to lay aside the spirit of contradiction, and lead them, perhaps, to do battle in their own persons for reasonings which they find considered demonstrations by all men of genius, lest they should seem not to understand them ; and, finally, the rest of mankind will readily trust to so many testimonies, and there will no longer be any one who will venture to doubt either the existence of God or the real distinction of mind and body. It is for you, in your singular wisdom, to judge of the importance of the establishment of such beliefs, [who are cognisant of the disorders which doubt of these truths produces].* But it would not here become me to commend at greater length the cause of God and of religion to you, who have always proved the strongest support of the Catholic Church.

* The *square* brackets, here and throughout the volume, are used to mark additions to the original of the revised French translation.

PREFACE TO THE READER.

I HAVE already slightly touched upon the questions respecting the existence of God and the nature of the human soul, in the "Discourse on the Method of rightly conducting the Reason, and seeking truth in the Sciences," published in French in the year 1637; not, however, with the design of there treating of them fully, but only, as it were, in passing, that I might learn from the judgments of my readers in what way I should afterwards handle them: for these questions appeared to me to be of such moment as to be worthy of being considered more than once, and the path which I follow in discussing them is so little trodden, and so remote from the ordinary route, that I thought it would not be expedient to illustrate it at greater length in French, and in a discourse that might be read by all, lest even the more feeble minds should believe that this path might be entered upon by them.

But, as in the Discourse on Method, I had requested all who might find aught meriting censure in my writings, to do me the favour of pointing it out to me, I may state that no objections worthy of remark have been alleged against what I then said on these questions, except two, to which I will here briefly reply, before undertaking their more detailed discussion.

The first objection is that though, while the human mind reflects on itself, it does not perceive[1]* that it is any

* See Note I. The numbers refer to the Notes, in which will be found some notices of the various terms throughout the volume that appeared to require a word of comment.

other than a thinking thing, it does not follow that its nature or essence consists only in its being a thing which thinks; so that the word *only* shall exclude all other things which might also perhaps be said to pertain to the nature of the mind.

To this objection I reply, that it was not my intention in that place to exclude these according to the order of truth in the matter (of which I did not then treat), but only according to the order of thought (perception); so that my meaning was, that I clearly apprehended nothing, so far as I was conscious, as belonging to my essence, except that I was a thinking thing, or a thing possessing in itself the faculty of thinking. But I will show hereafter how, from the consciousness that nothing besides thinking belongs to the essence of the mind, it follows that nothing else does in truth belong to it.

The second objection is that it does not follow, from my possessing the idea of a thing more perfect than I am, that the idea itself is more perfect than myself, and much less that what is represented by the idea exists.

But I reply that in the term *idea*[2] there is here something equivocal; for it may be taken either materially for an act of the understanding, and in this sense it cannot be said to be more perfect than I, or objectively, for the thing represented by that act, which, although it be not supposed to exist out of my understanding, may, nevertheless, be more perfect than myself, by reason of its essence. But, in the sequel of this treatise I will show more amply how, from my possessing the idea of a thing more perfect than myself, it follows that this thing really exists.

Besides these two objections, I have seen, indeed, two treatises of sufficient length relating to the present matter. In these, however, my conclusions, much more than my premises, were impugned, and that by arguments borrowed from the common places of the atheists. But, as arguments of this sort can make no impression on the minds of those who shall rightly understand my reasonings, and as the judgments of many are so irrational and weak that they are persuaded rather by the opinions on a subject that are first presented to them, however false and opposed to reason they may be, than by a true and solid, but subsequently received, refutation of them, I am unwilling here to reply to these strictures from a dread of being, in the first instance, obliged to state them.

I will only say, in general, that all which the atheists commonly allege in favour of the non-existence of God, arises continually from one or other of these two things, namely, either the ascription of human affections to Deity, or the undue attribution to our minds of so much vigour and wisdom that we may essay to determine and comprehend both what God can and ought to do; hence all that is alleged by them will occasion us no difficulty, provided only we keep in remembrance that our minds must be considered finite, while Deity is incomprehensible and infinite.

Now that I have once, in some measure, made proof of the opinions of men regarding my work, I again undertake to treat of God and the human soul, and at the same time to discuss the principles of the entire First Philosophy, without, however, expecting any commendation from the crowd for my endeavours, or a wide circle of readers. On the contrary, I would advise none to read this work, unless such as are able and willing to meditate with me in earnest, to detach their minds from commerce with the senses, and likewise to deliver themselves from all prejudice; and individuals of this character are, I well know, remarkably rare. But with regard to those who, without caring to comprehend the order and connection of the reasonings, shall study only detached clauses for the purpose of small but noisy criticism, as is the custom with many, I may say that such persons will not profit greatly by the reading of this treatise; and although perhaps they may find opportunity for cavilling in several places, they will yet hardly start any pressing objections, or such as shall be deserving of reply.

But since, indeed, I do not promise to satisfy others on all these subjects at first sight, nor arrogate so much to myself as to believe that I have been able to foresee all that may be the source of difficulty to each one, I shall expound, first of all, in the *Meditations*, those considerations by which I feel persuaded that I have arrived at a certain and evident knowledge of truth, in order that I may ascertain whether the reasonings which have prevailed with myself will also be effectual in convincing others. I will then reply to the objections of some men, illustrious for their genius and learning, to whom these Meditations were sent for criticism before they were committed to the press; for these objections are so numerous and varied that I venture to anticipate that nothing, at least nothing of any moment, will readily occur to any mind which has not been touched upon in them.

Hence it is that I earnestly entreat my readers not to come to any judgment on the questions raised in the Meditations until they have taken care to read the whole of the Objections, with the relative Replies.

SYNOPSIS

OF THE

SIX FOLLOWING MEDITATIONS.

In the First Meditation I expound the grounds on which we may doubt in general of all things, and especially of material objects, so long, at least, as we have no other foundations for the sciences than those we have hitherto possessed. Now, although the utility of a doubt so general may not be manifest at first sight, it is nevertheless of the greatest, since it delivers us from all prejudice, and affords the easiest pathway by which the mind may withdraw itself from the senses; and, finally, makes it impossible for us to doubt wherever we afterwards discover truth.

In the Second, the mind which, in the exercise of the freedom peculiar to itself, supposes that no object is, of the existence of which it has even the slightest doubt, finds that, meanwhile, it must itself exist. And this point is likewise of the highest moment, for the mind is thus enabled easily to distinguish what pertains to itself, that is, to the intellectual nature, from what is to be referred to the body. But since some, perhaps, will expect, at this stage of our progress, a statement of the reasons which establish the doctrine of the immortality of the soul, I think it proper here to make such aware, that it was my aim to write nothing of which I could not give exact demonstration, and that I therefore felt myself obliged to adopt an order similar to that in use among the geometers, viz., to premise all upon which the proposition in question depends, before coming to any conclusion respecting it. Now, the first and chief pre-requisite for the knowledge of the immortality of the soul is our being able to form the clearest possible conception (*conceptus*—concept) of the soul itself, and such as shall be absolutely distinct from all our notions of body; and how this is to be accomplished is there shown. There is required, besides this, the assurance that all objects which we clearly and distinctly think are true (really exist) in that very mode in which we think them; and this could not be established previously to the Fourth Meditation. Farther, it is necessary, for the same purpose, that we possess a distinct conception of corporeal nature, which is given partly in the Second and partly in the Fifth and Sixth Meditations. And, finally, on these grounds, we are necessitated to conclude, that all those objects which are clearly and distinctly conceived to be diverse substances, as mind and body, are substances really reciprocally distinct; and this inference is made in the Sixth Meditation. The absolute distinction of mind and body is, besides, confirmed in this Second Meditation, by showing that we cannot conceive body unless as divisible; while, on the other hand, mind cannot be conceived unless as indivisible. For we are not able to conceive the half of a mind, as we can of any body, however small, so that the natures of these two substances are to be held, not only as diverse, but even in some measure as contraries. I have not, however, pursued this discussion further in the present treatise, as well for the reason that these considerations are sufficient to show that the destruction of the mind does not follow from the corruption of the body, and thus to afford to men the hope of a future life, as also because the premises from which it is competent for us to infer the immortality of the soul, involve an explication of the whole principles of Physics: in order to establish, in the first place, that generally all substances, that is, all things which can exist only in consequence of having been created by God, are in their own nature incorruptible, and can never cease to be, unless God himself, by refusing his concurrence to them, reduce them to nothing; and, in the second place, that body, taken generally, is a substance, and therefore can never perish, but that the human body, in as far as it differs from other bodies, is constituted only by a certain configuration of members, and by other accidents of this sort, while the human mind is not made up of accidents, but is a pure substance. For although all the accidents of the mind be changed—although, for example, it think certain things, will others, and perceive others, the mind itself does not vary with these changes; while, on the contrary, the human body is no longer the same if a change take place in the form of any of its parts: from which it follows that the body may, indeed, without difficulty perish, but that the mind is in its own nature immortal.

In the Third Meditation, I have unfolded at sufficient length, as appears to me, my chief argument for the existence of God. But yet, since I was there desirous to avoid the use of comparisons taken from material objects, that I might withdraw, as far as possible, the minds of my readers from the senses, numerous obscurities perhaps remain, which, however, will, I trust, be afterwards entirely removed in the Replies to the Objections: thus, among other things, it may be difficult to understand how the idea of a being absolutely perfect, which is found in our minds, possesses so much objective reality[3] [*i. e.*, participates by representation in so many degrees of being and perfection] that it must be held to arise from a cause absolutely perfect. This is illustrated in the Replies by the comparison of a highly perfect machine, the idea of which exists in the mind of some workman; for as the objective (*i. e.*, representative) perfection of this idea must have some cause, viz., either the science of the workman, or of some other person from whom he has received the idea, in the same way the idea of God, which is found in us, demands God himself for its cause.

In the Fourth, it is shown that all which we clearly and distinctly perceive (apprehend) is true; and, at the same time, is explained wherein consists the nature of error; points that require to be known as well for confirming the preceding truths, as for the better understanding of those that are to follow. But, meanwhile, it must be observed, that I do not at all there treat of Sin, that is, of error committed in the pursuit of good and evil, but of that sort alone which arises in the determination of the true and the false. Nor do I refer to matters of faith, or to the conduct of life, but only to what regards speculative truths, and such as are known by means of the natural light alone.

In the Fifth, besides the illustration of corporeal nature, taken generically, a new demonstration is given of the existence of God, not free, perhaps, any more than the former, from certain difficulties, but of these the solution will be found in the Replies to the Objections. I further show, in what sense it is true that the certitude of geometrical demonstrations themselves is dependent on the knowledge of God.

Finally, in the Sixth, the act of the understanding (*intellectio*) is distinguished from that of the imagination (*imaginatio*); the marks of this distinction are described; the human mind is shown to be really distinct from the body, and, nevertheless, to be so closely conjoined therewith, as together to form, as it were, a unity. The whole of the errors which arise from the senses are brought under review, while the means of avoiding them are pointed out; and, finally, all the grounds are adduced from which the existence of material objects may be inferred; not, however, because I deemed them of great utility in establishing what they prove, viz., that there is in reality a world, that men are possessed of bodies, and the like, the truth of which no one of sound mind ever seriously doubted; but because, from a close consideration of them, it is perceived that they are neither so strong nor clear as the reasonings which conduct us to the knowledge of our mind and of God; so that the latter are, of all which come under human knowledge, the most certain and manifest—a conclusion which it was my single aim in these Meditations to establish; on which account I here omit mention of the various other questions which, in the course of the discussion, I had occasion likewise to consider.

MEDITATIONS

ON

THE FIRST PHILOSOPHY,

IN WHICH

THE EXISTENCE OF GOD, AND THE REAL DISTINCTION OF MIND AND BODY, ARE DEMONSTRATED.

MEDITATION I.

OF THE THINGS OF WHICH WE MAY DOUBT.

SEVERAL years have now elapsed since I first became aware that I had accepted, even from my youth, many false opinions for true, and that consequently what I afterwards based on such principles was highly doubtful; and from that time I was convinced of the necessity of undertaking once in my life to rid myself of all the opinions I had adopted, and of commencing anew the work of building from the foundation, if I desired to establish a firm and abiding superstructure in the sciences. But as this enterprise appeared to me to be one of great magnitude, I waited until I had attained an age so mature as to leave me no hope that at any stage of life more advanced I should be better able to execute my design. On this account, I have delayed so long that I should henceforth consider I was doing wrong were I still to consume in deliberation any of the time that now remains for action. To-day, then, since I have opportunely freed my mind from all cares, [and am happily disturbed by no passions], and since I am in the secure possession of leisure in a peaceable retirement, I will at length apply myself earnestly and freely to the general overthrow of all my former opinions. But, to this end, it will not be necessary for me to show that the whole of these are false—a point, perhaps, which I shall never reach; but as even now my reason convinces me that I ought not the less carefully to withhold belief from what is not entirely certain and indubitable, than from what is manifestly false, it will be sufficient to justify the rejection of the whole if I shall find in each some ground for doubt. Nor for this purpose will it be necessary even to deal with each belief individually, which would be truly an endless labour; but, as the removal from below of the foundation necessarily involves the downfall of the whole edifice, I will at once approach the criticism of the principles on which all my former beliefs rested.

All that I have, up to this moment, accepted as possessed of the highest truth and certainty, I received either from or through the senses.[4] I observed, however, that these sometimes misled us; and it is the part of prudence not to place absolute confidence in that by which we have even once been deceived.

But it may be said, perhaps, that, although the senses occasionally mislead us respecting minute objects, and such as are so far removed from us as to be beyond the reach of close observation, there are yet many other of their informations (presentations), of the truth of which it is manifestly impossible to doubt; as for example, that I am in this place, seated by the fire, clothed in a winter dressing-gown, that I hold in my hands this piece of paper, with other intimations of the same nature. But how could I deny that I possess these hands and this body, and withal escape being classed with persons in a state of insanity, whose brains are so disordered and clouded by dark bilious vapours as to cause them pertinaciously to assert that they are monarchs when they are in the greatest poverty; or clothed [in gold] and purple when destitute of any covering; or that their head is made of clay, their body of glass, or that they are gourds? I should certainly be not less insane than they, were I to regulate my procedure according to examples so extravagant.

Though this be true, I must nevertheless here consider that I am a man, and that, consequently, I am in the habit of sleeping, and representing to myself in dreams those same things, or even sometimes others less probable, which the insane think are presented to them in their waking moments. How often have I dreamt that I was in these familiar circumstances,—that I was dressed, and occupied this place by the fire, when I was lying undressed in bed? At the present moment, however, I certainly look upon this paper with eyes wide awake; the head which I now move is not asleep; I extend this hand consciously and with express purpose, and I perceive it; the occurrences in sleep are not so distinct as all this. But I cannot forget that, at other times, I have been deceived in sleep by similar illusions; and, attentively considering those cases, I perceive so clearly that there exist no certain marks by which the state of waking can ever be distinguished from sleep, that I feel greatly astonished; and in amazement I almost persuade myself that I am now dreaming.

Let us suppose, then, that we are dreaming, and that all these particulars—namely, the opening of the eyes, the motion of the head, the forth-putting of the hands—are merely illusions; and even that we really possess neither an entire body nor hands such as we see. Nevertheless, it must be admitted at least that the objects which appear to us in sleep are, as it were, painted representations which could not have been formed unless in the likeness of realities; and, therefore, that those general objects, at all events,—namely, eyes, a head, hands, and an entire body—are not simply imaginary, but really existent. For, in truth, painters themselves, even when they study to represent sirens and satyrs by forms the most fantastic and extraordinary, cannot bestow upon them natures absolutely new, but can only make a certain medley of the members of different animals; or if they chance to imagine something so novel that nothing at all similar has ever been seen before, and such as is, therefore, purely fictitious and absolutely false, it is at least certain that the colours of which this is composed are real.

And on the same principle, although these general objects, viz. [a body], eyes, a head, hands, and the like, be imaginary, we are nevertheless absolutely necessitated to admit the reality at least of some other objects still more simple and universal than these, of which, just as of certain real colours, all those images of things, whether true and real, or false and fantastic, that are found in our consciousness (*cogitatio*),[5] are formed.

To this class of objects seem to belong corporeal nature in general and its extension; the figure of extended things, their quantity or magnitude, and their number, as also the place in, and the time during, which they exist, and other things of the same sort. We will not, therefore, perhaps reason illegitimately if we conclude from this that Physics, Astronomy, Medicine, and all the other sciences that have for their end the consideration of composite objects, are indeed of a doubtful character; but that Arithmetic, Geometry, and the other sciences of the same class, which regard merely the simplest and most general objects, and scarcely inquire whether or not these are really existent, contain somewhat that is certain and indubitable: for whether I am awake or dreaming, it remains true that two and three make five, and that a square has but four sides; nor does it seem possible that truths so apparent can ever fall under a suspicion of falsity [or incertitude].

Nevertheless, the belief that there is a God who is all-powerful, and who created me, such as I am, has, for a long time, obtained steady possession of my mind. How, then, do I know that he has not arranged that there should be neither earth, nor sky, nor any extended thing, nor figure, nor magnitude, nor place, providing at the same time, however, for [the rise in me of the perceptions of all these objects, and] the persuasion that these do not exist otherwise than as I perceive them? And further, as I sometimes think that others are in error respecting matters of which they believe themselves to possess a perfect knowledge, how do I know that I am not also deceived each time I add together two and three, or number the sides of a square, or form some judgment still more simple, if more simple indeed can be imagined? But perhaps Deity has not been willing that I should be thus deceived, for He is said to be supremely good. If, however, it were repugnant to the goodness of Deity to have created me subject to constant deception, it would seem likewise to be contrary to his goodness to allow me to be occasionally deceived; and yet it is clear that this is permitted. Some, indeed, might perhaps be found who would be disposed rather to deny the existence of a Being so powerful than to believe that there is nothing certain. But let us for the present refrain from opposing this opinion, and grant that all which is here said of a Deity is fabulous: nevertheless, in whatever way it be supposed that I reached the state in which I exist, whether by fate, or chance, or by an endless series of antecedents and consequents, or by any other means, it is clear (since to be deceived and to err is a certain defect) that the probability of my being so imperfect as to be the constant victim of deception. will be increased exactly in proportion as the power possessed by the cause, to which they assign my origin, is lessened. To these reasonings I have assuredly nothing to reply, but am constrained at last to avow that there is nothing of all that I formerly believed to be true of which it is impossible to doubt, and that not through thoughtlessness or levity, but from cogent and maturely considered reasons; so that henceforward, if I desire to discover anything certain, I ought not the less carefully to refrain from assenting to those same opinions than to what might be shown to be manifestly false.

But it is not sufficient to have made these observations; care must be taken likewise to keep them in remembrance. For those old and customary opinions perpetually recur—long and familiar usage giving them the right of occupying my mind, even almost against my will, and subduing my belief; nor will I lose the habit of deferring to them and confiding in them so long as I shall consider them to be what in truth they are, viz., opinions to some extent doubtful, as I have already shown, but still highly probable, and such as it is much more reasonable to believe than deny. It is for this reason I am persuaded that I shall not be doing wrong, if, taking an opposite judgment of deliberate design, I become my own deceiver, by supposing, for a time, that all those opinions are entirely false and imaginary, until at length, having thus balanced my old by my new prejudices, my judgment shall no longer be turned aside by perverted usage from the path that may conduct to the perception of truth. For I am assured that, meanwhile, there will arise neither peril nor error from this course, and that I cannot for the present yield too much to distrust, since the end I now seek is not action but knowledge.

I will suppose, then, not that Deity, who is sovereignly good and the fountain of truth, but that some malignant demon, who is at once exceedingly potent and deceitful, has employed all his artifice to deceive me; I will suppose that the sky, the air, the earth, colours, figures, sounds, and all external things, are nothing better than the illusions of dreams, by means of which this being has laid snares for my credulity; I will consider myself as without hands, eyes, flesh, blood, or any of the senses, and as falsely believing that I am possessed of these; I will continue resolutely fixed in this belief, and if indeed by this means it be not in my power to arrive at the knowledge of truth, I shall at least do what is in my power, viz., [suspend my judgment], and guard with settled purpose against giving my assent to what is false, and being imposed upon by this deceiver, whatever be his power and artifice.

But this undertaking is arduous, and a certain indolence insensibly leads me back to my ordinary course of life; and just as the captive, who, perchance, was enjoying in his dreams an imaginary liberty, when he begins to suspect that it is but a vision, dreads awakening, and conspires with the agreeable illusions that the deception may be prolonged; so I, of my own accord, fall back into the train of my former beliefs, and fear to arouse myself from my slumber, lest the time of laborious wakefulness that would succeed this quiet rest, in place of bringing any light of day, should prove inadequate to dispel the darkness that will arise from the difficulties that have now been raised.

MEDITATION II.

OF THE NATURE OF THE HUMAN MIND; AND THAT IT IS MORE EASILY KNOWN THAN THE BODY.

The Meditation of yesterday has filled my mind with so many doubts, that it is no longer in my power to forget them. Nor do I see, meanwhile, any principle on which they can be resolved; and, just as if I had fallen all of a sudden into very deep water, I am so greatly disconcerted as to be unable either to plant my feet firmly on the bottom or sustain myself by swimming on the surface. I will, nevertheless, make an effort, and try anew the same path on which I had entered yesterday, that is, proceed by casting aside all that admits of the slightest doubt, not less than if I had discovered it to be absolutely false; and I will continue always in this track until I shall find something that is certain, or at least, if I can do nothing more, until I shall know with certainty that there is nothing certain. Archimedes, that he might transport the entire globe from the place it occupied to another, demanded only a point that was firm and immoveable; so also, I shall be entitled to entertain the highest expectations, if I am fortunate enough to discover only one thing that is certain and indubitable.

I suppose, accordingly, that all the things which I see are false (fictitious); I believe that none of those objects which my fallacious memory represents ever existed; I suppose that I possess no senses; I believe that body, figure, extension, motion, and place are merely fictions of my mind. What is there, then, that can be esteemed true? Perhaps this only, that there is absolutely nothing certain.

But how do I know that there is not something different altogether from the objects I have now enumerated, of which it is impossible to entertain the slightest doubt? Is there not a God, or some being, by whatever name I may designate him, who causes these thoughts to arise in my mind? But why suppose such a being, for it may be I myself am capable of producing them? Am I, then, at least not something? But I before denied that I possessed senses or a body; I hesitate, however, for what follows from that? Am I so dependent on the body and the senses that without these I cannot exist? But I had the persuasion that there was absolutely nothing in the world, that there was no sky and no earth, neither minds nor bodies; was I not, therefore, at the same time, persuaded that I did not exist? Far from it; I assuredly existed, since I was persuaded. But there is I know not what being, who is possessed at once of the highest power and the deepest cunning, who is constantly employing all his ingenuity in deceiving me. Doubtless, then, I exist, since I am deceived; and, let him deceive me as he may, he can never bring it about that I am nothing, so long as I shall be conscious that I am something. So that it must, in fine, be maintained, all things being maturely and carefully considered, that this proposition (*pronunciatum*) I am, I exist, is necessarily true each time it is expressed by me, or conceived in my mind.

But I do not yet know with sufficient clearness what I am, though assured that I am; and hence, in the next place, I must take care, lest perchance I inconsiderately substitute some other object in room of what is properly myself, and thus wander from truth, even in that knowledge (cognition) which I hold to be of all others the most certain and evident. For this reason, I will now consider anew what I formerly believed myself to be, before I entered on the present train of thought; and of my previous opinion I will retrench all that can in the least be invalidated by the grounds of doubt I have adduced, in order that there may at length remain nothing but what is certain and indubitable. What then did I formerly think I was? Undoubtedly I judged that I was a man. But what is a man? Shall I say a rational animal? Assuredly not; for it would be necessary forthwith to inquire into what is meant by animal, and what by rational, and thus, from a single question, I should insensibly glide into others, and these more difficult than the first; nor do I now possess enough of leisure to warrant me in wasting my time amid subtleties of this sort. I prefer here to attend to the thoughts that sprung up of themselves in my mind, and were inspired by my own nature alone, when I applied myself to the consideration of what I was. In the first place, then, I thought that I possessed a countenance, hands, arms, and all the fabric of members that appears in a corpse, and which I called by the name of body. It further occurred to me that I was nourished, that I walked, perceived, and thought, and all those actions I referred to the soul; but what the soul itself was I either did not stay to consider, or, if I did, I imagined that it was something extremely rare and subtile, like wind, or flame, or ether, spread through my grosser parts. As regarded the body, I did not even doubt of its nature, but thought I distinctly knew it, and if I had wished to describe it according to the notions I then entertained, I should have explained myself in this manner: By body I understand all that can be terminated by a certain figure; that can be comprised in a certain place, and so fill a certain space as therefrom to exclude every other body; that can be perceived either by touch, sight, hearing, taste, or smell; that can be moved in different ways, not indeed of itself, but by something foreign to it by which it is touched [and from which it receives the impression]; for the power of self-motion, as likewise that of perceiving and thinking, I held as by no means pertaining to the nature of body; on the contrary, I was somewhat astonished to find such faculties existing in some bodies.

But [as to myself, what can I now say that I am], since I suppose there exists an extremely powerful, and, if I may so speak, malignant being, whose whole endeavours are directed towards deceiving me? Can I affirm that I possess any one of all those attributes of which I have lately spoken as belonging to the nature of body? After attentively considering them in my own mind, I find none of them that can properly be said to belong to myself. To recount them were idle and tedious. Let us pass, then, to the attributes of the soul. The first mentioned were the powers of nutrition and walking; but, if it be true that I have no body, it is true likewise that I am capable neither of walking nor of being nourished. Perception is another attribute of the soul; but perception too is impossible without the body: besides, I have frequently, during sleep, believed that I perceived objects which I afterwards observed I did not in reality perceive. Thinking is another attribute of the soul; and here I discover what properly belongs to myself. This alone is inseparable from me. I am—I exist: this is certain; but how often? As often as I think; for perhaps it would even happen, if I should wholly cease to think, that I should at the same time altogether cease to be. I now admit nothing that is not necessarily true: I am therefore, precisely speaking, only a thinking thing, that is, a mind (*mens sive animus*), understanding, or reason,—terms whose signification was before unknown to me. I am, however, a real thing, and really existent; but what thing? The answer was, a thinking thing. The question now arises, am I aught besides? I will stimulate my imagination with a view to discover whether I am not still something more than a thinking being. Now it is plain I am not the assemblage of members called the human body; I am not a thin and penetrating air diffused through all these members, or wind, or flame, or vapour, or breath, or any of all the things I can imagine; for I supposed that all these were not, and, without changing the supposition, I find that I still feel assured of my existence.

But it is true, perhaps, that those very things which I suppose to be non-existent, because they are unknown to me, are not in truth different from myself whom I know. This is a point I cannot determine, and do not now enter into any dispute regarding it. I can only judge of things that are known to me: I am conscious that I exist, and I who know that I exist inquire into what I am. It is, however, perfectly certain that the knowledge of my existence, thus precisely taken, is not dependent on things, the existence of which is as yet unknown to me: and consequently it is not dependent on any of the things I can feign in imagination. Moreover, the phrase itself, I frame an image (*effingo*), reminds me of my error; for I should in truth frame one if I were to imagine myself to be anything, since to imagine is nothing more than to contemplate the figure or image of a corporeal thing; but I already know that I exist, and that it is possible at the same time that all those images, and in general all that relates to the nature of body, are merely dreams [or chimeras]. From this I discover that it is not more reasonable to say, I will excite my imagination that I may know more distinctly what I am, than to express myself as follows: I am now awake, and perceive something real; but because my perception is not sufficiently clear, I will of express purpose go to sleep that my dreams may represent to me the object of my perception with more truth and clearness. And, therefore, I know that nothing of all that I can embrace in imagination belongs to the knowledge which I have of myself, and that there is need to recall with the utmost care the mind from this mode of thinking, that it may be able to know its own nature with perfect distinctness.

But what, then, am I? A thinking thing, it has been said. But what is a thinking thing? It is a thing that doubts, understands, [conceives], affirms, denies, wills, refuses, that imagines also, and perceives. Assuredly it is not little, if all these properties belong to my nature. But why should they not belong to it? Am I not that very being who now doubts of almost everything; who, for all that, understands and conceives certain things; who affirms one alone as true, and denies the others; who desires to know more of them, and does not wish to be deceived; who imagines many things, sometimes even despite his will; and is likewise percipient of many, as if through the medium of the senses. Is there nothing of all this as true as that I am, even although I should be always dreaming, and although he who gave me being employed all his ingenuity to deceive me? Is there also any one of these attributes that can be properly distinguished from my thought, or that can be said to be separate from myself? For it is of itself so evident that it is I who doubt, I who understand, and I who desire, that it is here unnecessary to add anything by way of rendering it more clear. And I am as certainly the same being who imagines; for, although it may be (as I before supposed) that nothing I imagine is true, still the power of imagination does not cease really to exist in me and to form part of my thought. In fine, I am the same being who perceives, that is, who apprehends certain objects as by the organs of sense, since, in truth, I see light, hear a noise, and feel heat. But it will be said that these presentations are false, and that I am dreaming. Let it be so. At all events it is certain that I seem to see light, hear a noise, and feel heat; this cannot be false, and this is what in me is properly called perceiving (*sentire*), which is nothing else than thinking. From this I begin to know what I am with somewhat greater clearness and distinctness than heretofore.

But, nevertheless, it still seems to me, and I cannot help believing, that corporeal things, whose images are formed by thought, [which fall under the senses], and are examined by the same, are known with much greater distinctness than that I know not what part of myself which is not imaginable; although, in truth, it may seem strange to say that I know and comprehend with greater distinctness things whose existence appears to me doubtful, that are unknown, and do not belong to me, than others of whose reality I am persuaded, that are known to me, and appertain to my proper nature; in a word, than myself. But I see clearly what is the state of the case. My mind is apt to wander, and will not yet submit to be restrained within the limits of truth. Let us therefore leave the mind to itself once more, and, according to it every kind of liberty, [permit it to consider the objects that appear to it from without], in order that, having afterwards withdrawn it from these gently and opportunely, [and fixed it on the consideration of its being and the properties it finds in itself], it may then be the more easily controlled.

Let us now accordingly consider the objects that are commonly thought to be [the most easily, and likewise] the most distinctly known, viz., the bodies we touch and see; not, indeed, bodies in general, for these general notions are usually somewhat more confused, but one body in particular. Take, for example, this piece of wax; it is quite fresh, having been but recently taken from the beehive; it has not yet lost the sweetness of the honey it contained; it still retains somewhat of the odour of the flowers from which it was gathered; its colour, figure, size, are apparent (to the sight); it is hard, cold, easily handled; and sounds when struck upon with the finger. In fine, all that contributes to make a body as distinctly known as possible, is found in the one before us. But, while I am speaking, let it be placed near the fire—what remained of the taste exhales, the smell evaporates, the colour changes, its figure is destroyed, its size increases, it becomes liquid, it grows hot, it can hardly be handled, and, although struck upon, it emits no sound. Does the same wax still remain after this change? It must be admitted that it does remain ; no one doubts it, or judges otherwise. What, then, was it I knew with so much distinctness in the piece of wax? Assuredly, it could be nothing of all that I observed by means of the senses, since all the things that fell under taste, smell, sight, touch, and hearing are changed, and yet the same wax remains. It was perhaps what I now think, viz., that this wax was neither the sweetness of honey, the pleasant odour of flowers, the whiteness, the figure, nor the sound, but only a body that a little before appeared to me conspicuous under these forms, and which is now perceived under others. But, to speak precisely, what is it that I imagine when I think of it in this way? Let it be attentively considered, and, retrenching all that does not belong to the wax, let us see what remains. There certainly remains nothing, except something extended, flexible, and moveable. But what is meant by flexible and moveable? Is it not that I imagine that the piece of wax, being round, is capable of becoming square, or of passing from a square into a triangular figure? Assuredly such is not the case, because I conceive that it admits of an infinity of similar changes ; and I am, moreover, unable to compass this infinity by imagination, and consequently this conception which I have of the wax is not the product of the faculty of imagination. But what now is this extension? Is it not also unknown? for it becomes greater when the wax is melted, greater when it is boiled, and greater still when the heat increases ; and I should not conceive [clearly and] according to truth, the wax as it is, if I did not suppose that the piece we are considering admitted even of a wider variety of extension than I ever imagined. I must, therefore, admit that I cannot even comprehend by imagination what the piece of wax is, and that it is the mind alone (*mens*, Lat., *entendement*, F.) which perceives it. I speak of one piece in particular; for, as to wax in general, this is still more evident. But what is the piece of wax that can be perceived only by the [understanding or] mind? It is certainly the same which I see, touch, imagine; and, in fine, it is the same which, from the beginning, I believed it to be. But (and this it is of moment to observe) the perception of it is neither an act of sight, of touch, nor of imagination, and never was either of these, though it might formerly seem so, but is simply an intuition (*inspectio*) of the mind, which may be imperfect and confused, as it formerly was, or very clear and distinct, as it is at present, according as the attention is more or less directed to the elements which it contains, and of which it is composed.

But, meanwhile, I feel greatly astonished when I observe [the weakness of my mind, and] its proneness to error. For although, without at all giving expression to what I think, I consider all this in my own mind, words yet occasionally impede my progress, and I am almost led into error by the terms of ordinary language. We say, for example, that we see the same wax when it is before us, and not that we judge it to be the same from its retaining the same colour and figure: whence I should forthwith be disposed to conclude that the wax is known by the act of sight, and not by the intuition of the mind alone, were it not for the analogous instance of human beings passing on in the street below, as observed from a window. In this case I do not fail to say that I see the men themselves, just as I say that I see the wax; and yet what do I see from the window beyond hats and cloaks that might cover artificial machines, whose motions might be determined by springs? But I judge that there are human beings from these appearances, and thus I comprehend, by the faculty of judgment alone which is in the mind, what I believed I saw with my eyes.

The man who makes it his aim to rise to knowledge superior to the common, ought to be ashamed to seek occasions of doubting from the vulgar forms of speech: instead, therefore, of doing this, I shall proceed with the matter in hand, and inquire whether I had a clearer and more perfect perception of the piece of wax when I first saw it, and when I thought I knew it by means of the external sense itself, or, at all events, by the common sense (*sensus communis*), as it is called, that is, by the imaginative faculty; or whether I rather apprehend it more clearly at present, after having examined with greater care, both what it is, and in what way it can be known. It would certainly be ridiculous to entertain any doubt on this point. For what, in that first perception, was there distinct? What did I perceive which any animal might not have perceived? But when I distinguish the wax from its exterior forms, and when, as if I had stripped it of its vestments, I consider it quite naked, it is certain, although some error may still be found in my judgment, that I cannot, nevertheless, thus apprehend it without possessing a human mind.

But, finally, what shall I say of the mind itself, that is, of myself? for as yet I do not admit that I am anything but mind. What, then! I who seem to possess so distinct an apprehension of the piece of wax,—do I not know myself, both with greater truth and certitude, and also much more distinctly and clearly? For if I judge that the wax exists because I see it, it assuredly follows, much more evidently, that I myself am or exist, for the same reason: for it is possible that what I see may not in truth be wax, and that I do not even possess eyes with which to see anything; but it cannot be that when I see, or, which comes to the same thing, when I think I see, I myself who think am nothing. So likewise, if I judge that the wax exists because I touch it, it will still also follow that I am ; and if I determine that my imagination, or any other cause, whatever it be, persuades me of the existence of the wax, I will still draw the same conclusion. And what is here remarked of the piece of wax, is applicable to all the other things that are external to me. And further, if the [notion or] perception of wax appeared to me more precise and distinct, after that not only sight and touch, but many other causes besides, rendered it manifest to my apprehension, with how much greater distinctness must I now know myself, since all the reasons that contribute to the knowledge of the nature of wax, or of any body whatever, manifest still better the nature of my mind? And there are besides so many other things in the mind itself that contribute to the illustration of its nature, that those dependent on the body, to which I have here referred, scarcely merit to be taken into account.

But, in conclusion, I find I have insensibly reverted to the point I desired; for, since it is now manifest to me that bodies themselves are not properly perceived by the senses nor by the faculty of imagination, but by the intellect alone; and since they are not perceived because they are seen and touched, but only because they are understood [or rightly comprehended by thought], I readily discover that there is nothing more easily or clearly apprehended than my own mind. But because it is difficult to rid one's self so promptly of an opinion to which one has been long accustomed, it will be desirable to tarry for some time at this stage, that, by long continued meditation, I may more deeply impress upon my memory this new knowledge.

MEDITATION III.

OF GOD: THAT HE EXISTS.

I WILL now close my eyes, I will stop my ears, I will turn away my senses from their objects, I will even efface from my consciousness all the images of corporeal things; or at least, because this can hardly be accomplished, I will consider them as empty and false; and thus, holding converse only with myself, and closely examining my nature, I will endeavour to obtain by degrees a more intimate and familiar knowledge of myself. I am a thinking (conscious) thing, that is, a being who doubts, affirms, denies, knows a few objects, and is ignorant of many,—[who loves, hates], wills, refuses,—who imagines likewise, and perceives; for, as I before remarked, although the things which I perceive or imagine are perhaps nothing at all apart from me [and in themselves], I am nevertheless assured that those modes of consciousness which I call perceptions and imaginations, in as far only as they are modes of consciousness, exist in me. And in the little I have said I think I have summed up all that I really know, or at least all that up to this time I was aware I knew. Now, as I am endeavouring to extend my knowledge more widely, I will use circumspection, and consider with care whether I can still discover in myself anything further which I have not yet hitherto observed. I am certain that I am a thinking thing; but do I not therefore likewise know what is required to render me certain of a truth? In this first knowledge, doubtless, there is nothing that gives me assurance of its truth except the clear and distinct perception of what I affirm, which would not indeed be sufficient to give me the assurance that what I say is true, if it could ever happen that anything I thus clearly and distinctly perceived should prove false; and accordingly it seems to me that I may now take as a general rule, that all that is very clearly and distinctly apprehended (conceived) is true.

Nevertheless I before received and admitted many things as wholly certain and manifest, which yet I afterwards found to be doubtful. What, then, were those? They were the earth, the sky, the stars, and all the other objects which I was in the habit of perceiving by the senses. But what was it that I clearly [and distinctly] perceived in them? Nothing more than that the ideas and the thoughts of those objects were presented to my mind. And even now I do not deny that these ideas are found in my mind. But there was yet another thing which I affirmed, and which, from having been accustomed to believe it, I thought I clearly perceived, although, in truth, I did not perceive it at all; I mean the existence of objects external to me, from which those ideas proceeded, and to which they had a perfect resemblance; and it was here I was mistaken, or if I judged correctly, this assuredly was not to be traced to any knowledge I possessed (the force of my perception, Lat.).

But when I considered any matter in arithmetic and geometry, that was very simple and easy, as, for example, that two and three added together make five, and things of this sort, did I not view them with at least sufficient clearness to warrant me in affirming their truth? Indeed, if I afterwards judged that we ought to doubt of these things, it was for no other reason than because it occurred to me that a God might perhaps have given me such a nature as that I should be deceived, even respecting the matters that appeared to me the most evidently true. But as often as this preconceived opinion of the sovereign power of a God presents itself to my mind, I am constrained to admit that it is easy for him, if he wishes it, to cause me to err, even in matters where I think I possess the highest evidence; and, on the other hand, as often as I direct my attention to things which I think I apprehend with great clearness, I am so persuaded of their truth that I naturally break out into expressions such as these: Deceive me who may, no one will yet ever be able to bring it about that I am not, so long as I shall be conscious that I am, or at any future time cause it to be true that I have never been, it being now true that I am, or make two and three more or less than five, in supposing which, and other like absurdities, I discover a manifest contradiction.

And in truth, as I have no ground for believing that Deity is deceitful, and as, indeed, I have not even considered the reasons by which the existence of a Deity of any kind is established, the ground of doubt that rests only on this supposition is very slight, and, so to speak, metaphysical. But, that I may be able wholly to remove it, I must inquire whether there is a God, as soon as an opportunity of doing so shall present itself; and if I find that there is a God, I must examine likewise whether he can be a deceiver; for, without the knowledge of these two truths, I do not see that I can ever be certain of anything. And that I may be enabled to examine this without interrupting the order of meditation I have proposed to myself [which is, to pass by degrees from the notions that I shall find first in my mind to those I shall afterwards discover in it], it is necessary at this stage to divide all my thoughts into certain classes, and to consider in which of these classes truth and error are, strictly speaking, to be found.

Of my thoughts some are, as it were, images of things, and to these alone properly belongs the name *idea;* as when I think [represent to my mind] a man, a chimera, the sky, an angel, or God. Others, again, have certain other forms; as when I will, fear, affirm, or deny, I always, indeed, apprehend something as the object of my thought, but I also embrace in thought something more than the representation of the object; and of this class of thoughts some are called volitions or affections, and others judgments.

Now, with respect to ideas, if these are considered only in themselves, and are not referred to any object beyond them, they cannot, properly speaking, be false; for, whether I imagine a goat or a chimera, it is not less true that I imagine the one than the other. Nor need we fear that falsity may exist in the will or affections; for, although I may desire objects that are wrong, and even that never existed, it is still true that I desire them. There thus only remain our judgments, in which we must take diligent heed that we be not deceived. But the chief and most ordinary error that arises in them consists in judging that the ideas which are in us are like or conformed to the things that are external to us; for assuredly, if we but considered the ideas themselves as certain modes of our thought (consciousness), without referring them to anything beyond, they would hardly afford any occasion of error.

But, among these ideas, some appear to me to be innate,[6] others adventitious, and others to be made by myself (factitious); for, as I have the power of conceiving what is called a thing, or a truth, or a thought, it seems to me that I hold this power from no other source than my own nature; but if I now hear a noise, if I see the sun, or if I feel heat, I have all along judged that these sensations proceeded from certain objects existing out of myself; and, in fine, it appears to me that sirens, hippogryphs, and the like, are inventions of my own mind. But I may even perhaps come to be of opinion that all my ideas are of the class which I call adventitious, or that they are all innate, or that they are all factitious, for I have not yet clearly discovered their true origin; and what I have here principally to do is to consider, with reference to those that appear to come from certain objects without me, what grounds there are for thinking them like these objects.

The first of these grounds is that it seems to me I am so taught by nature; and the second that I am conscious that those ideas are not dependent on my will, and therefore not on myself, for they are frequently presented to me against my will,—as at present, whether I will or not, I feel heat; and I am thus persuaded that this sensation or idea *(sensum vel ideam)* of heat is produced in me by something different from myself, viz., by the heat of the fire by which I sit. And it is very reasonable to suppose that this object impresses me with its own likeness rather than any other thing.

But I must consider whether these reasons are sufficiently strong and convincing. When I speak of being taught by nature in this matter, I understand by the word nature only a certain spontaneous impetus that impels me to believe in a resemblance between ideas and their objects, and not a natural light that affords a knowledge of its truth. But these two things are widely different; for what the natural light shows to be true can be in no degree doubtful, as, for example, that I am because I doubt, and other truths of the like kind: inasmuch as I possess no other faculty whereby to distinguish truth from error, which can teach me the falsity of what the natural light declares to be true, and which is equally trust-worthy; but with respect to [seemingly] natural impulses, I have observed, when the question related to the choice of right or wrong in action, that they frequently led me to take the worse part; nor do I see that I have any better ground for following them in what relates to truth and error. Then, with respect to the other reason, which is that because these ideas do not depend on my will, they must arise from objects existing without me, I do not find it more convincing than the former; for, just as those natural impulses, of which I have lately spoken, are found in me, notwithstanding that they are not always in harmony with my will, so likewise it may be that I possess some power not sufficiently known to myself capable of producing ideas without the aid of external objects, and, indeed, it has always hitherto appeared to me that they are formed during sleep, by some power of this nature, without the aid of aught external. And, in fine, although I should grant that they proceeded from those objects, it is not a necessary consequence that they must be like them. On the contrary, I have observed, in a number of instances, that there was a great difference between the object and its idea. Thus, for example, I find in my mind two wholly diverse ideas of the sun; the one, by which it appears to me extremely small, draws its origin from the senses, and should be placed in the class of adventitious ideas; the other, by which it seems to be many times larger than the whole earth, is taken up on astronomical grounds, that is, elicited from certain notions born with me, or is framed by myself in some other manner. These two ideas cannot certainly both resemble the same sun; and reason teaches me that the one which seems to have immediately emanated from it is the most unlike. And these things sufficiently prove that hitherto it has not been from a certain and deliberate judgment, but only from a sort of blind impulse, that I believed in the existence of certain things different from myself, which, by the organs of sense, or by whatever other means it might be, conveyed their ideas or images into my mind [and impressed it with their likenesses].

But there is still another way of inquiring whether, of the objects whose ideas are in my mind, there are any that exist out of me. If ideas are taken in so far only as they are certain modes of consciousness, I do not remark any difference or inequality among them, and all seem, in the same manner, to proceed from myself; but, considering them as images, of which one represents one thing and another a different, it is evident that a great diversity obtains among them. For, without doubt, those that represent substances are something more, and contain in themselves, so to speak, more objective reality [that is, participate by representation in higher degrees of being or perfection], than those that represent only modes or accidents; and again, the idea by which I conceive a God [sovereign], eternal, infinite, [immutable], all-knowing, all-powerful, and the creator of all things that are out of himself,—this, I say, has certainly in it more objective reality than those ideas by which finite substances are represented.

Now, it is manifest by the natural light that there must at least be as much reality in the efficient and total cause as in its effect; for whence can the effect draw its reality if not from its cause? and how could the cause communicate to it this reality unless it possessed it in itself? And hence it follows, not only that what is cannot be produced by what is not, but likewise that the more perfect,—in other words, that which contains in itself more reality,—cannot be the effect of the less perfect: and this is not only evidently true of those effects, whose reality is actual or formal, but likewise of ideas, whose reality is only considered as objective. Thus, for example, the stone that is not yet in existence, not only cannot now commence to be, unless it be produced by that which possesses in itself, formally or eminently,[7] all that enters into its composition, [in other words, by that which contains in itself the same properties that are in the stone, or others superior to them]; and heat can only be produced in a subject that was before devoid of it, by a cause that is of an order, [degree or kind], at least as perfect as heat; and so of the others. But further, even the idea of the heat, or of the stone, cannot exist in me unless it be put there by a cause that contains, at least, as much reality as I conceive existent in the heat or in the stone: for, although that cause may not transmit into my idea anything of its actual or formal reality, we ought not on this account to imagine that it is less real; but we ought to consider that, [as every idea is a work of the mind], its nature is such as of itself to demand no other formal reality than that which it borrows from our consciousness, of which it is but a mode, [that is, a manner or way of thinking]. But in order that an idea may contain this objective reality rather than that, it must doubtless derive it from some cause in which is found at least as much formal reality as the idea contains of objective; for, if we suppose that there is found in an idea anything which was not in its cause, it must of course derive this from nothing. But, however imperfect may be the mode of existence by which a thing is objectively [or by representation] in the understanding by its idea, we certainly cannot, for all that, allege that this mode of existence is nothing, nor, consequently, that the idea owes its origin to nothing. Nor must it be imagined that, since the reality which is considered in these ideas is only objective, the same reality need not be formally (actually) in the causes of these ideas, but only objectively: for, just as the mode of existing objectively belongs to ideas by their peculiar nature, so likewise the mode of existing formally appertains to the causes of these ideas (at least to the first and principal), by their peculiar nature. And although an idea may give rise to another idea, this regress cannot, nevertheless, be infinite; we must in the end reach a first idea, the cause of which is, as it were, the archetype in which all the reality [or perfection] that is found objectively [or by representation] in these ideas is contained formally [and in act]. I am thus clearly taught by the natural light that ideas exist in me as pictures or images. which may in truth readily fall short of the perfection of the objects from which they are taken, but can never contain anything greater or more perfect.

And in proportion to the time and care with which I examine all those matters, the conviction of their truth brightens and becomes distinct. But, to sum up, what conclusion shall I draw from it all? It is this;—if the objective reality [or perfection] of any one of my ideas be such as clearly to convince me, that this same reality exists in me neither formally nor eminently, and if, as follows from this, I myself cannot be the cause of it, it is a necessary consequence that I am not alone in the world, but that there is besides myself some other being who exists as the cause of that idea; while, on the contrary, if no such idea be found in my mind, I shall have no sufficient ground of assurance of the existence of any other being besides myself; for, after a most careful search, I have, up to this moment, been unable to discover any other ground.

But, among these my ideas, besides that which represents myself, respecting which there can be here no difficulty, there is one that represents a God; others that represent corporeal and inanimate things; others angels; others animals; and, finally, there are some that represent men like myself. But with respect to the ideas that represent other men, or animals, or angels, I can easily suppose that they were formed by the mingling and composition of the other ideas which I have of myself, of corporeal things, and of God, although there were, apart from myself, neither men, animals, nor angels. And with regard to the ideas of corporeal objects, I never discovered in them anything so great or excellent which I myself did not appear capable of originating; for, by considering these ideas closely and scrutinising them individually, in the same way that I yesterday examined the idea of wax, I find that there is but little in them that is clearly and distinctly perceived. As belonging to the class of things that are clearly apprehended, I recognise the following, viz., magnitude or extension in length, breadth, and depth; figure, which results from the termination of extension; situation, which bodies of diverse figures preserve with reference to each other; and motion or the change of situation; to which may be added substance, duration, and number. But with regard to light, colours, sounds, odours, tastes, heat, cold, and the other tactile qualities, they are thought with so much obscurity and confusion, that I cannot determine even whether they are true or false; in other words, whether or not the ideas I have of these qualities are in truth the ideas of real objects. For although I before remarked that it is only in judgments that formal falsity, or falsity properly so called, can be met with, there may nevertheless be found in ideas a certain material falsity, which arises when they represent what is nothing as if it were something. Thus, for example, the ideas I have of cold and heat are so far from being clear and distinct, that I am unable from them to discover whether cold is only the privation of heat, or heat the privation of cold; or whether they are or are not real qualities: and since, ideas being as it were images, there can be none that does not seem to us to represent some object, the idea which represents cold as something real and positive will not improperly be called false, if it be correct to say that cold is nothing but a privation of heat; and so in other cases. To ideas of this kind, indeed, it is not necessary that I should assign any author besides myself: for if they are false, that is, represent objects that are unreal, the natural light teaches me that they proceed from nothing; in other words, that they are in me only because something is wanting to the perfection of my nature; but if these ideas are true, yet because they exhibit to me so little reality that I cannot even distinguish the object represented from non-being, I do not see why I should not be the author of them.

With reference to those ideas of corporeal things that are clear and distinct, there are some which, as appears to me, might have been taken from the idea I have of myself, as those of substance, duration, number, and the like. For when I think that a stone is a substance, or a thing capable of existing of itself, and that I am likewise a substance, although I conceive that I am a thinking and non-extended thing, and that the stone, on the contrary, is extended and unconscious, there being thus the greatest diversity between the two concepts,—yet these two ideas seem to have this in common that they both represent substances. In the same way, when I think of myself as now existing, and recollect besides that I existed some time ago, and when I am conscious of various thoughts whose number I know, I then acquire the ideas of duration and number, which I can afterwards transfer to as many objects as I please. With respect to the other qualities that go to make up the ideas of corporeal objects, viz., extension, figure, situation, and motion, it is true that they are not formally in me, since I am merely a thinking being; but because they are only certain modes of substance, and because I myself am a substance, it seems possible that they may be contained in me eminently.

There only remains, therefore, the idea of God, in which I must consider whether there is anything that cannot be supposed to originate with myself. By the name God, I understand a substance infinite, [eternal, immutable], independent, all-knowing, all-powerful, and by which I myself, and every other thing that exists, if any such there be, were created. But these properties are so great and excellent, that the more attentively I consider them the less I feel persuaded that the idea I have of them owes its origin to myself alone. And thus it is absolutely necessary to conclude, from all that I have before said, that God exists: for though the idea of substance be in my mind owing to this, that I myself am a substance, I should not, however, have the idea of an infinite substance, seeing I am a finite being, unless it were given me by some substance in reality infinite.

And I must not imagine that I do not apprehend the infinite by a true idea, but only by the negation of the finite, in the same way that I comprehend repose and darkness by the negation of motion and light: since, on the contrary, I clearly perceive that there is more reality in the infinite substance than in the finite, and therefore that in some way I possess the perception (notion) of the infinite before that of the finite, that is, the perception of God before that of myself, for how could I know that I doubt, desire, or that something is wanting to me, and that I am not wholly perfect, if I possessed no idea of a being more perfect than myself, by comparison of which I knew the deficiencies of my nature?

And it cannot be said that this idea of God is perhaps materially false, and consequently that it may have arisen from nothing, [in other words, that it may exist in me from my imperfection], as I before said of the ideas of heat and cold, and the like: for, on the contrary, as this idea is very clear and distinct, and contains in itself more objective reality than any other, there can be no one of itself more true, or less open to the suspicion of falsity.

The idea, I say, of a being supremely perfect, and infinite, is in the highest degree true; for although, perhaps, we may imagine that such a being does not exist, we cannot, nevertheless, suppose that his idea represents nothing real, as I have already said of the idea of cold. It is likewise clear and distinct in the highest degree, since whatever the mind clearly and distinctly conceives as real or true, and as implying any perfection, is contained entire in this idea. And this is true, nevertheless, although I do not comprehend the infinite, and although there may be in God an infinity of things that I cannot comprehend, nor perhaps even compass by thought in any way; for it is of the nature of the infinite that it should not be comprehended by the finite; and it is enough that I rightly understand this, and judge that all which I clearly perceive, and in which I know there is some perfection, and perhaps also an infinity of properties of which I am ignorant, are formally or eminently in God, in order that the idea I have of him may become the most true, clear, and distinct of all the ideas in my mind.

But perhaps I am something more than I suppose myself to be, and it may be that all those perfections which I attribute to God, in some way exist potentially in me, although they do not yet show themselves, and are not reduced to act. Indeed, I am already conscious that my knowledge is being increased [and perfected] by degrees; and I see nothing to prevent it from thus gradually increasing to infinity, nor any reason why, after such increase and perfection, I should not be able thereby to acquire all the other perfections of the Divine nature; nor, in fine, why the power I possess of acquiring those perfections, if it really now exist in me, should not be sufficient to produce the ideas of them. Yet, on looking more closely into the matter, I discover that this cannot be; for, in the first place, although it were true that my knowledge daily acquired new degrees of perfection, and although there were potentially in my nature much that was not as yet actually in it, still all these excellences make not the slightest approach to the idea I have of the Deity, in whom there is no perfection merely potentially [but all actually] existent; for it is even an unmistakeable token of imperfection in my knowledge, that it is augmented by degrees. Further, although my knowledge increase more and more, nevertheless I am not, therefore, induced to think that it will ever be actually infinite, since it can never reach that point beyond which it shall be incapable of further increase. But I conceive God as actually infinite, so that nothing can be added to his perfection. And, in fine, I readily perceive that the objective being of an idea cannot be produced by a being that is merely potentially existent, which, properly speaking, is nothing, but only by a being existing formally or actually.

And, truly, I see nothing in all that I have now said which it is not easy for any one, who shall carefully consider it, to discern by the natural light; but when I allow my attention in some degree to relax, the vision of my mind being obscured, and, as it were, blinded by the images of sensible objects, I do not readily remember the reason why the idea of a being more perfect than myself, must of necessity have proceeded from a being in reality more perfect. On this account I am here desirous to inquire further, whether I, who possess this idea of God, could exist supposing there were no God. And I ask, from whom could I, in that case, derive my existence? Perhaps from myself, or from my parents, or from some other causes less perfect than God; for anything more perfect, or even equal to God, cannot be thought or imagined. But if I [were independent of every other existence, and] were myself the author of my being, I should doubt of nothing, I should desire nothing, and, in fine, no perfection would be awanting to me; for I should have bestowed upon myself every perfection of which I possess the idea, and I should thus be God. And it must not be imagined that what is now wanting to me is perhaps of more difficult acquisition than that of which I am already possessed; for, on the contrary, it is quite manifest that it was a matter of much higher difficulty that I, a thinking being, should arise from nothing, than it would be for me to acquire the knowledge of many things of which I am ignorant, and which are merely the accidents of a thinking substance; and certainly, if I possessed of myself the greater perfection of which I have now spoken, [in other words, if I were the author of my own existence], I would not at least have denied to myself things that may be more easily obtained, [as that infinite variety of knowledge of which

I am at present destitute]. I could not, indeed, have denied to myself any property which I perceive is contained in the idea of God, because there is none of these that seems to me to be more difficult to make or acquire; and if there were any that should happen to be more difficult to acquire, they would certainly appear so to me (supposing that I myself were the source of the other things I possess), because I should discover in them a limit to my power. And though I were to suppose that I always was as I now am, I should not, on this ground, escape the force of these reasonings, since it would not follow, even on this supposition, that no author of my existence needed to be sought after. For the whole time of my life may be divided into an infinity of parts, each of which is in no way dependent on any other; and, accordingly, because I was in existence a short time ago, it does not follow that I must now exist, unless in this moment some cause create me anew as it were,—that is, conserve me. In truth, it is perfectly clear and evident to all who will attentively consider the nature of duration, that the conservation of a substance, in each moment of its duration, requires the same power and act that would be necessary to create it, supposing it were not yet in existence; so that it is manifestly a dictate of the natural light that conservation and creation differ merely in respect of our mode of thinking [and not in reality]. All that is here required, therefore, is that I interrogate myself to discover whether I possess any power by means of which I can bring it about that I, who now am, shall exist a moment afterwards: for, since I am merely a thinking thing (or since, at least, the precise question, in the meantime, is only of that part of myself), if such a power resided in me, I should, without doubt, be conscious of it; but I am conscious of no such power, and thereby I manifestly know that I am dependent upon some being different from myself.

But perhaps the being upon whom I am dependent, is not God, and I have been produced either by my parents, or by some causes less perfect than Deity. This cannot be: for, as I before said, it is perfectly evident that there must at least be as much reality in the cause as in its effect; and accordingly, since I am a thinking thing, and possess in myself an idea of God, whatever in the end be the cause of my existence, it must of necessity be admitted that it is likewise a thinking being, and that it possesses in itself the idea and all the perfections I attribute to Deity. Then it may again be inquired whether this cause owes its origin and existence to itself, or to some other cause. For if it be self-existent, it follows, from what I have before laid down, that this cause is God; for, since it possesses the perfection of self-existence, it must likewise, without doubt, have the power of actually possessing every perfection of which it has the idea,—in other words, all the perfections I conceive to belong to God. But if it owe its existence to another cause than itself, we demand again, for a similar reason, whether this second cause exists of itself or through some other, until, from stage to stage, we at length arrive at an ultimate cause, which will be God. And it is quite manifest that in this matter there can be no infinite regress of causes, seeing that the question raised respects not so much the cause which once produced me, as that by which I am at this present moment conserved.

Nor can it be supposed that several causes concurred in my production, and that from one I received the idea of one of the perfections I attribute to Deity, and from another the idea of some other, and thus that all those perfections are indeed found somewhere in the universe, but do not all exist together in a single being who is God; for, on the contrary, the unity, the simplicity or inseparability of all the properties of Deity, is one of the chief perfections I conceive him to possess; and the idea of this unity of all the perfections of Deity could certainly not be put into my mind by any cause from which I did not likewise receive the ideas of all the other perfections; for no power could enable me to embrace them in an inseparable unity, without at the same time giving me the knowledge of what they were [and of their existence in a particular mode].

Finally, with regard to my parents [from whom it appears I sprung], although all that I believed respecting them be true, it does not, nevertheless, follow that I am conserved by them, or even that I was produced by them, in so far as I am a thinking being. All that, at the most, they contributed to my origin was the giving of certain dispositions (modifications) to the matter in which I have hitherto judged that I or my mind, which is what alone I now consider to be myself, is enclosed; and thus there can here be no difficulty with respect to them, and it is absolutely necessary to conclude from this alone that I am, and possess the idea of a being absolutely perfect, that is, of God, that his existence is most clearly demonstrated.

There remains only the inquiry as to the way in which I received this idea from God; for I have not drawn it from the senses, nor is it even presented to me unexpectedly, as is usual with the ideas of sensible objects, when these are presented or appear to be presented to the external organs of the senses; it is not even a pure production or fiction of my mind, for it is not in my power to take from or add to it; and consequently there but remains the alternative that it is innate, in the same way as is the idea of myself. And, in truth, it is not to be wondered at that God, at my creation, implanted this idea in me, that it might serve, as it were, for the mark of the workman impressed on his work; and it is not also necessary that the mark should be something different from the work itself; but considering only that God is my creator, it is highly probable that he in some way fashioned me after his own image and likeness, and that I perceive this likeness, in which is contained the idea of God, by the same faculty by which I apprehend myself,—in other words, when I make myself the object of reflection, I not only find that I am an incomplete, [imperfect] and dependent being, and one who unceasingly aspires after something better and greater than he is; but, at the same time, I am assured likewise that he upon whom I am dependent possesses in himself all the goods after which I aspire, [and the ideas of which I find in my mind], and that not merely indefinitely and potentially, but infinitely and actually, and that he is thus God. And the whole force of the argument of which I have here availed myself to establish the existence of God, consists in this, that I perceive I could not possibly be of such a nature as I am, and yet have in my mind the idea of a God, if God did not in reality exist,—this same God, I say, whose idea is in my mind—that is, a being who possesses all those lofty perfections, of which the mind may have some slight conception, without, however, being able fully to comprehend them,—and who is wholly superior to all defect, [and has nothing that marks imperfection]: whence it is sufficiently manifest that he cannot be a deceiver, since it is a dictate of the natural light that all fraud and deception spring from some defect.

But before I examine this with more attention, and pass on to the consideration of other truths that may be evolved out of it, I think it proper to remain here for some time in the contemplation of God himself—that I may ponder at leisure his marvellous attributes—and behold, admire, and adore the beauty of this light so unspeakably great, as far, at least, as the strength of my mind, which is to some degree dazzled by the sight, will permit. For just as we learn by faith that the supreme felicity of another life consists in the contemplation of the Divine majesty alone, so even now we learn from experience that a like meditation, though incomparably less perfect, is the source of the highest satisfaction of which we are susceptible in this life.

MEDITATION IV.

OF TRUTH AND ERROR.

I HAVE been habituated these bygone days to detach my mind from the senses, and I have accurately observed that there is exceedingly little which is known with certainty respecting corporeal objects,—that we know much more of the human mind, and still more of God himself. I am thus able now without difficulty to abstract my mind from the contemplation of [sensible or] imaginable objects, and apply it to those which, as disengaged from all matter, are purely intelligible. And certainly the idea I have of the human mind in so far as it is a thinking thing, and not extended in length, breadth, and depth, and participating in none of the properties of body, is incomparably more distinct than the idea of any corporeal object; and when I consider that I doubt, in other words, that I am an incomplete and dependent being, the idea of a complete and independent being, that is to say of God, occurs to my mind with so much clearness and distinctness,—and from the fact alone that this idea is found in me, or that I who possess it exist, the conclusions that God exists, and that my own existence, each moment of its continuance, is absolutely dependent upon him, are so manifest,—as to lead me to believe it impossible that the human mind can know anything with more clearness and certitude. And now I seem to discover a path that will conduct us from the contemplation of the true God, in whom are contained all the treasures of science and wisdom, to the knowledge of the other things in the universe.

For, in the first place, I discover that it is impossible for him ever to deceive me, for in all fraud and deceit there is a certain imperfection: and although it may seem that the ability to deceive is a mark of subtlety or power, yet the will testifies without doubt of malice and weakness; and such, accordingly, cannot be found in God. In the next place, I am conscious that I possess a certain faculty of judging [or discerning truth from error], which I doubtless received from God, along with whatever else is mine; and since it is impossible that he should will to deceive me, it is likewise certain that he has not given me a faculty that will ever lead me into error, provided I use it aright.

And there would remain no doubt on this head, did it not seem to follow from this, that I can never therefore be deceived; for if all I possess be from God, and if he planted in me no faculty that is deceitful, it seems to follow that I can never fall into error. Accordingly, it is true that when I think only of God (when I look upon myself as coming from God, Fr.), and turn wholly to him, I discover [in myself] no cause of error or falsity: but immediately thereafter, recurring to myself, experience assures me that I am nevertheless subject to innumerable errors. When I come to inquire into the cause of these, I observe that there is not only present to my consciousness a real and positive idea of God, or of a being supremely perfect, but also, so to speak, a certain negative idea of nothing,—in other words, of that which is at an infinite distance from every sort of perfection, and that I am, as it were, a mean between God and nothing, or placed in such a way between absolute existence and non-existence, that there is in truth nothing in me to lead me into error, in so far as an absolute being is my creator; but that, on the other hand, as I thus likewise participate in some degree of nothing or of non-being, in other words, as I am not myself the supreme Being, and as I am wanting in many perfections, it is not surprising I should fall into error. And I hence discern that error, so far as error is not something real, which depends for its existence on God, but is simply defect; and therefore that, in order to fall into it, it is not necessary God should have given me a faculty expressly for this end, but that my being deceived arises from the circumstance that the power which God has given me of discerning truth from error is not infinite.

Nevertheless this is not yet quite satisfactory; for error is not a pure negation, [in other words, it is not the simple deficiency or want of some knowledge which is not due], but the privation or want of some knowledge which it would seem I ought to possess. But, on considering the nature of God, it seems impossible that he should have planted in his creature any faculty not perfect in its kind, that is, wanting in some perfection due to it: for if it be true, that in proportion to the skill of the maker the perfection of his work is greater, what thing can have been produced by the supreme Creator of the universe that is not absolutely perfect in all its parts? And assuredly there is no doubt that God could have created me such as that I should never be deceived; it is certain, likewise, that he always wills what is best: is it better, then, that I should be capable of being deceived than that I should not?

Considering this more attentively, the first thing that occurs to me is the reflection that I must not be surprised if I am not always capable of comprehending the reasons why God acts as he does; nor must I doubt of his existence because I find, perhaps, that there are several other things besides the present respecting which I understand neither why nor how they were created by him; for, knowing already that my nature is extremely weak and limited, and that the nature of God, on the other hand, is immense, incomprehensible, and infinite, I have no longer any difficulty in discerning that there is an infinity of things in his power whose causes transcend the grasp of my mind: and this consideration alone is sufficient to convince me, that the whole class of final causes is of no avail in physical [or natural] things; for it appears to me that I cannot, without exposing myself to the charge of temerity, seek to discover the [impenetrable] ends of Deity.

It further occurs to me that we must not consider only one creature apart from the others, if we wish to determine the perfection of the works of Deity, but generally all his creatures together; for the same object that might perhaps, with some show of reason, be deemed highly imperfect if it were alone in the world, may for all that be the most perfect possible, considered as forming part of the whole universe: and although, as it was my purpose to doubt of everything, I only as yet know with certainty my own existence and that of God, nevertheless, after having remarked the infinite power of Deity, I cannot deny that he may have produced many other objects, or at least that he is able to produce them, so that I may occupy a place in the relation of a part to the great whole of his creatures.

Whereupon, regarding myself more closely, and considering what my errors are (which alone testify to the existence of imperfection in me), I observe that these depend on the concurrence of two causes, viz., the faculty of cognition which I possess, and that of election or the power of free choice,—in other words, the understanding and the will. For by the understanding alone, I [neither affirm nor deny anything, but] merely apprehend (*percipio*) the ideas regarding which I may form a judgment; nor is any error, properly so called, found in it thus accurately taken. And although there are perhaps innumerable objects in the world of which I have no idea in my understanding, it cannot, on that account, be said that I am deprived of those ideas [as of something that is due to my nature], but simply that I do not possess them, because, in truth, there is no ground to prove that Deity ought to have endowed me with a larger faculty of cognition than he has actually bestowed upon me; and however skilful a workman I suppose him to be, I have no reason, on that account, to think that it was obligatory on him to give to each of his works all the perfections he is able to bestow upon some. Nor, moreover, can I complain that God has not given me freedom of choice, or a will sufficiently ample and perfect, since, in truth, I am conscious of will so ample and extended as to be superior to all limits. And what appears to me here to be highly remarkable is that, of all the other properties I possess, there is none so great and perfect as that I do not clearly discern it could be still greater and more perfect. For, to take an example, if I consider the faculty of understanding which I possess, I find that it is of very small extent, and greatly limited, and at the same time I form the idea of another faculty of the same nature, much more ample and even infinite; and seeing that I can frame the idea of it, I discover, from this circumstance alone, that it pertains to the nature of God. In the same way, if I examine the faculty of memory or imagination, or any other faculty I possess, I find none that is not small and circumscribed, and in God immense [and infinite]. It is the faculty of will only, or freedom of choice, which I experience to be so great that I am unable to conceive the idea of another that shall be more ample and extended; so that it is chiefly my will which leads me to discern that I bear a certain image and similitude of Deity. For although the faculty of will is incomparably greater in God than in myself, as well in respect of the knowledge and power that are conjoined with it, and that render it stronger and more efficacious, as in respect of the object, since in him it extends to a greater number of things, it does not, nevertheless, appear to me greater, considered in itself formally and precisely: for the power of will consists only in this, that we are able to do or not to do the same thing (that is, to affirm or deny, to pursue or shun it), or rather in this alone, that in affirming or denying, pursuing or shunning, what is proposed to us by the understanding, we so act that we are not conscious of being determined to a particular action by any external force. For, to the possession of freedom, it is not necessary that I be alike indifferent towards each of two contraries; but, on the contrary, the more I am inclined towards the one, whether because I clearly know that in it there is the reason of truth and goodness, or because God thus internally disposes my thought, the more freely do I choose and embrace it; and assuredly divine grace and natural knowledge, very far from diminishing liberty, rather augment and fortify it. But the indifference of which I am conscious when I am not impelled to one side rather than to another for want of a reason, is the lowest grade of liberty, and manifests defect or negation of knowledge rather than perfection of will; for if I always clearly knew what was true and good, I should never have any difficulty in determining what judgment I ought to come to, and what choice I ought to make, and I should thus be entirely free without ever being indifferent.

From all this I discover, however, that neither the power of willing, which I have received from God, is of itself the source of my errors, for it is exceedingly ample and perfect in its kind; nor even the power of understanding, for as I conceive no object unless by means of the faculty that God bestowed upon me, all that I conceive is doubtless rightly conceived by me, and it is impossible for me to be deceived in it.

Whence, then, spring my errors? They arise from this cause alone, that I do not restrain the will, which is of much wider range than the understanding, within the same limits, but extend it even to things I do not understand,

and as the will is of itself indifferent to such, it readily falls into error and sin by choosing the false in room of the true, and evil instead of good.

For example, when I lately considered whether aught really existed in the world, and found that because I considered this question, it very manifestly followed that I myself existed, I could not but judge that what I so clearly conceived was true, not that I was forced to this judgment by any external cause, but simply because great clearness of the understanding was succeeded by strong inclination in the will; and I believed this the more freely and spontaneously in proportion as I was less indifferent with respect to it. But now I not only know that I exist, in so far as I am a thinking being, but there is likewise presented to my mind a certain idea of corporeal nature; hence I am in doubt as to whether the thinking nature which is in me, or rather which I myself am, is different from that corporeal nature, or whether both are merely one and the same thing, and I here suppose that I am as yet ignorant of any reason that would determine me to adopt the one belief in preference to the other: whence it happens that it is a matter of perfect indifference to me which of the two suppositions I affirm or deny, or whether I form any judgment at all in the matter.

This indifference, moreover, extends not only to things of which the understanding has no knowledge at all, but in general also to all those which it does not discover with perfect clearness at the moment the will is deliberating upon them; for, however probable the conjectures may be that dispose me to form a judgment in a particular matter, the simple knowledge that these are merely conjectures, and not certain and indubitable reasons, is sufficient to lead me to form one that is directly the opposite. Of this I lately had abundant experience, when I laid aside as false all that I had before held for true, on the single ground that I could in some degree doubt of it. But if I abstain from judging of a thing when I do not conceive it with sufficient clearness and distinctness, it is plain that I act rightly, and am not deceived; but if I resolve to deny or affirm, I then do not make a right use of my free will; and if I affirm what is false, it is evident that I am deceived: moreover, even although I judge according to truth, I stumble upon it by chance, and do not therefore escape the imputation of a wrong use of my freedom; for it is a dictate of the natural light, that the knowledge of the understanding ought always to precede the determination of the will.

And it is this wrong use of the freedom of the will in which is found the privation that constitutes the form of error. Privation, I say, is found in the act, in so far as it proceeds from myself, but it does not exist in the faculty which I received from God, nor even in the act, in so far as it depends on him; for I have assuredly no reason to complain that God has not given me a greater power of intelligence or more perfect natural light than he has actually bestowed, since it is of the nature of a finite understanding not to comprehend many things, and of the nature of a created understanding to be finite; on the contrary, I have every reason to render thanks to God, who owed me nothing, for having given me all the perfections I possess, and I should be far from thinking that he has unjustly deprived me of, or kept back, the other perfections which he has not bestowed upon me.

I have no reason, moreover, to complain because he has given me a will more ample than my understanding, since, as the will consists only of a single element, and that indivisible, it would appear that this faculty is of such a nature that nothing could be taken from it [without destroying it]; and certainly, the more extensive it is, the more cause I have to thank the goodness of him who bestowed it upon me.

And, finally, I ought not also to complain that God concurs with me in forming the acts of this will, or the judgments in which I am deceived, because those acts are wholly true and good, in so far as they depend on God; and the ability to form them is a higher degree of perfection in my nature than the want of it would be. With regard to privation, in which alone consists the formal reason of error and sin, this does not require the concurrence of Deity, because it is not a thing [or existence], and if it be referred to God as to its cause, it ought not to be called privation, but negation, [according to the signification of these words in the schools.] For in truth it is no imperfection in Deity that he has accorded to me the power of giving or withholding my assent from certain things of which he has not put a clear and distinct knowledge in my understanding; but it is doubtless an imperfection in me that I do not use my freedom aright, and readily give my judgment on matters which I only obscurely and confusedly conceive.

I perceive, nevertheless, that it was easy for Deity so to have constituted me as that I should never be deceived, although I still remained free and possessed of a limited knowledge, viz., by implanting in my understanding a clear and distinct knowledge of all the objects respecting which I should ever have to deliberate; or simply by so deeply engraving on my memory the resolution to judge of nothing without previously possessing a clear and distinct conception of it, that I should never forget it. And I easily understand that, in so far as I consider myself as a single whole, without reference to any other being in the universe, I should have been much more perfect than I now am, had Deity created me superior to error; but I cannot therefore deny that it is not somehow a greater perfection in the universe, that certain of its parts are not exempt from defect, as others are, than if they were all perfectly alike.

And I have no right to complain because God, who placed me in the world, was not willing that I should sustain that character which of all others is the chief and most perfect; I have even good reason to remain satisfied on the ground that, if he has not given me the perfection of being superior to error by the first means I have pointed out above, which depends on a clear and evident knowledge of all the matters regarding which I can deliberate, he has at least left in my power the other means, which is, firmly to retain the resolution never to judge where the truth is not clearly known to me: for, although I am conscious of the weakness of not being able to keep my mind continually fixed on the same thought, I can nevertheless, by attentive and oft-repeated meditation, impress it so strongly on my memory that I shall never fail to recollect it as often as I require it, and I can acquire in this way the habitude of not erring; and since it is in being superior to error that the highest and chief perfection of man consists, I deem that I have not gained little by this day's meditation, in having discovered the source of error and falsity.

And certainly this can be no other than what I have now explained: for as often as I so restrain my will within the limits of my knowledge, that it forms no judgment except regarding objects which are clearly and distinctly represented to it by the understanding, I can never be deceived; because every clear and distinct conception is doubtless something, and as such cannot owe its origin to nothing, but must of necessity have God for its author—God, I say, who, as supremely perfect, cannot, without a contradiction, be the cause of any error; and consequently it is necessary to conclude that every such conception [or judgment] is true. Nor have I merely learned to-day what I must avoid to escape error, but also what I must do to arrive at the knowledge of truth; for I will assuredly reach truth if I only fix my attention sufficiently on all the things I conceive perfectly, and separate these from others which I conceive more confusedly and obscurely: to which for the future I shall give diligent heed.

MEDITATION V.

OF THE ESSENCE OF MATERIAL THINGS; AND, AGAIN, OF GOD; THAT HE EXISTS.

SEVERAL other questions remain for consideration respecting the attributes of God and my own nature or mind. I will, however, on some other occasion perhaps resume the investigation of these. Meanwhile, as I have discovered what must be done and what avoided to arrive at the knowledge of truth, what I have chiefly to do is to essay to emerge from the state of doubt in which I have for some time been, and to discover whether anything can be known with certainty regarding material objects. But before considering whether such objects as I conceive exist without me, I must examine their ideas in so far as these are to be found in my consciousness, and discover which of them are distinct and which confused.

In the first place, I distinctly imagine that quantity which the philosophers commonly call continuous, or the extension in length, breadth, and depth that is in this quantity, or rather in the object to which it is attributed. Further, I can enumerate in it many diverse parts, and attribute to each of these all sorts of sizes, figures, situations, and local motions; and, in fine, I can assign to each of these motions all degrees of duration. And I not only distinctly know these things when I thus consider them in general; but besides, by a little attention, I discover innumerable particulars respecting figures, numbers, motion, and the like, which are so evidently true, and so accordant with my nature, that when I now discover them I do not so much appear to learn anything new, as to call to remembrance what I before knew, or for the first time to remark what was before in my mind, but to which I had not hitherto directed my attention. And what I here find of most importance is, that I discover in my mind innumerable ideas of certain objects, which cannot be esteemed pure negations, although perhaps they possess no reality beyond my thought, and which are not framed by me though it may be in my power to think, or not to think them, but possess true and immutable natures of their own. As, for example, when I imagine a triangle, although there is not perhaps and never was in any place in the universe apart from my thought one such figure, it remains true nevertheless that this figure possesses a certain determinate nature, form, or essence, which is immutable and eternal, and not framed by me, nor in any degree dependent on my thought; as appears from the circumstance, that diverse properties of the triangle may be demonstrated, viz., that its three angles are equal to two right, that its greatest side is subtended by its greatest angle, and the like, which, whether I will or not, I now clearly discern to belong to it, although before I did not at all think of them, when, for the first time, I imagined a triangle, and which accordingly cannot be said to have been invented by me. Nor is it a valid objection to allege, that perhaps this idea of a triangle came into my mind by the medium of the senses, through my having seen bodies of a triangular figure; for I am able to form in thought an innumerable variety of figures with regard to which it cannot be supposed that they were ever objects of sense, and I can nevertheless demonstrate diverse properties of their nature no less than of the triangle, all of which are assuredly true since I clearly conceive them: and they are therefore something, and not mere negations; for it is highly evident that all that is true is something, [truth being identical with existence];

and I have already fully shown the truth of the principle, that whatever is clearly and distinctly known is true. And although this had not been demonstrated, yet the nature of my mind is such as to compel me to assent to what I clearly conceive while I so conceive it; and I recollect that even when I still strongly adhered to the objects of sense, I reckoned among the number of the most certain truths those I clearly conceived relating to figures, numbers, and other matters that pertain to arithmetic and geometry, and in general to the pure mathematics.

But now if because I can draw from my thought the idea of an object, it follows that all I clearly and distinctly apprehend to pertain to this object, does in truth belong to it, may I not from this derive an argument for the existence of God? It is certain that I no less find the idea of a God in my consciousness, that is, the idea of a being supremely perfect, than that of any figure or number whatever: and I know with not less clearness and distinctness that an [actual and] eternal existence pertains to his nature than that all which is demonstrable of any figure or number really belongs to the nature of that figure or number; and, therefore, although all the conclusions of the preceding Meditations were false, the existence of God would pass with me for a truth at least as certain as I ever judged any truth of mathematics to be, although indeed such a doctrine may at first sight appear to contain more sophistry than truth. For, as I have been accustomed in every other matter to distinguish between existence and essence, I easily believe that the existence can be separated from the essence of God, and that thus God may be conceived as not actually existing. But, nevertheless, when I think of it more attentively, it appears that the existence can no more be separated from the essence of God, than the idea of a mountain from that of a valley, or the equality of its three angles to two right angles, from the essence of a [rectilineal] triangle; so that it is not less impossible to conceive a God, that is, a being supremely perfect, to whom existence is awanting, or who is devoid of a certain perfection, than to conceive a mountain without a valley.

But though, in truth, I cannot conceive a God unless as existing, any more than I can a mountain without a valley, yet, just as it does not follow that there is any mountain in the world merely because I conceive a mountain with a valley, so likewise, though I conceive God as existing, it does not seem to follow on that account that God exists; for my thought imposes no necessity on things; and as I may imagine a winged horse, though there be none such, so I could perhaps attribute existence to God, though no God existed. But the cases are not analogous, and a fallacy lurks under the semblance of this objection: for because I cannot conceive a mountain without a valley, it does not follow that there is any mountain or valley in existence, but simply that the mountain or valley, whether they do or do not exist, are inseparable from each other; whereas, on the other hand, because I cannot conceive God unless as existing, it follows that existence is inseparable from him, and therefore that he really exists: not that this is brought about by my thought, or that it imposes any necessity on things, but, on the contrary, the necessity which lies in the thing itself, that is, the necessity of the existence of God, determines me to think in this way: for it is not in my power to conceive a God without existence, that is, a being supremely perfect, and yet devoid of an absolute perfection, as I am free to imagine a horse with or without wings.

Nor must it be alleged here as an objection, that it is in truth necessary to admit that God exists, after having supposed him to possess all perfections, since existence is one of them, but that my original supposition was not necessary; just as it is not necessary to think that all quadrilateral figures can be inscribed in the circle, since, if I supposed this, I should be constrained to admit that the rhombus, being a figure of four sides, can be therein inscribed, which, however, is manifestly false. This objection is, I say, incompetent; for although it may not be necessary that I shall at any time entertain the notion of Deity, yet each time I happen to think of a first and sovereign being, and to draw, so to speak, the idea of him from the store-house of the mind, I am necessitated to attribute to him all kinds of perfections, though I may not then enumerate them all, nor think of each of them in particular. And this necessity is sufficient, as soon as I discover that existence is a perfection, to cause me to infer the existence of this first and sovereign being: just as it is not necessary that I should ever imagine any triangle, but whenever I am desirous of considering a rectilineal figure composed of only three angles, it is absolutely necessary to attribute those properties to it from which it is correctly inferred that its three angles are not greater than two right angles, although perhaps I may not then advert to this relation in particular. But when I consider what figures are capable of being inscribed in the circle, it is by no means necessary to hold that all quadrilateral figures are of this number; on the contrary, I cannot even imagine such to be the case, so long as I shall be unwilling to accept in thought aught that I do not clearly and distinctly conceive: and consequently there is a vast difference between false suppositions, as is the one in question, and the true ideas that were born with me, the first and chief of which is the idea of God. For indeed I discern on many grounds that this idea is not factitious, depending simply on my thought, but that it is the representation of a true and immutable nature: in the first place, because I can conceive no other being, except God, to whose essence existence [necessarily] pertains; in the second, because it is impossible to conceive two or more gods of this kind; and it being supposed that one such God exists, I clearly see that he must have existed from all eternity, and will exist to all eternity; and finally, because I apprehend many other properties in God, none of which I can either diminish or change.

But, indeed, whatever mode of probation I in the end adopt, it always returns to this, that it is only the things I clearly and distinctly conceive which have the power of completely persuading me. And although, of the objects I conceive in this manner, some, indeed, are obvious to every one, while others are only discovered after close and careful investigation; nevertheless, after they are once discovered, the latter are not esteemed less certain than the former. Thus, for example, to take the case of a right-angled triangle, although it is not so manifest at first that the square of the base is equal to the squares of the other two sides, as that the base is opposite to the greatest angle; nevertheless, after it is once apprehended, we are as firmly persuaded of the truth of the former as of the latter. And, with respect to God, if I were not pre-occupied by prejudices, and my thought beset on all sides by the continual presence of the images of sensible objects, I should know nothing sooner or more easily than the fact of his being. For is there any truth more clear than the existence of a Supreme Being, or of God, seeing it is to his essence alone that [necessary and eternal] existence pertains? And although the right conception of this truth has cost me much close thinking, nevertheless at present I feel not only as assured of it as of what I deem most certain, but I remark further that the certitude of all other truths is so absolutely dependent on it, that without this knowledge it is impossible ever to know anything perfectly.

For although I am of such a nature as to be unable, while I possess a very clear and distinct apprehension of a matter, to resist the conviction of its truth, yet because my constitution is also such as to incapacitate me from keeping my mind continually fixed on the same object, and as I frequently recollect a past judgment without at the same time being able to recall the grounds of it, it may happen meanwhile that other reasons are presented to me which would readily cause me to change my opinion, if I did not know that God existed; and thus I should possess no true and certain knowledge, but merely vague and vacillating opinions. Thus, for example, when I consider the nature of the [rectilineal] triangle, it most clearly appears to me, who have been instructed in the principles of geometry, that its three angles are equal to two right angles, and I find it impossible to believe otherwise, while I apply my mind to the demonstration; but as soon as I cease from attending to the process of proof, although I still remember that I had a clear comprehension of it, yet I may readily come to doubt of the truth demonstrated, if I do not know that there is a God: for I may persuade myself that I have been so constituted by nature as to be sometimes deceived, even in matters which I think I apprehend with the greatest evidence and certitude, especially when I recollect that I frequently considered many things to be true and certain which other reasons afterwards constrained me to reckon as wholly false.

But after I have discovered that God exists, seeing I also at the same time observed that all things depend on him, and that he is no deceiver, and thence inferred that all which I clearly and distinctly perceive is of necessity true: although I no longer attend to the grounds of a judgment, no opposite reason can be alleged sufficient to lead me to doubt of its truth, provided only I remember that I once possessed a clear and distinct comprehension of it. My knowledge of it thus becomes true and certain. And this same knowledge extends likewise to whatever I remember to have formerly demonstrated, as the truths of geometry and the like: for what can be alleged against them to lead me to doubt of them? Will it be that my nature is such that I may be frequently deceived? But I already know that I cannot be deceived in judgments of the grounds of which I possess a clear knowledge. Will it be that I formerly deemed things to be true and certain which I afterwards discovered to be false? But I had no clear and distinct knowledge of any of those things, and, being as yet ignorant of the rule by which I am assured of the truth of a judgment, I was led to give my assent to them on grounds which I afterwards discovered were less strong than at the time I imagined them to be. What further objection, then, is there? Will it be said that perhaps I am dreaming (an objection I lately myself raised), or that all the thoughts of which I am now conscious have no more truth than the reveries of my dreams? But although, in truth, I should be dreaming, the rule still holds that all which is clearly presented to my intellect is indisputably true.

And thus I very clearly see that the certitude and truth of all science depends on the knowledge alone of the true God, insomuch that, before I knew him, I could have no perfect knowledge of any other thing. And now that I know him, I possess the means of acquiring a perfect knowledge respecting innumerable matters, as well relative to God himself and other intellectual objects as to corporeal nature, in so far as it is the object of pure mathematics [which do not consider whether it exists or not].

MEDITATION VI.

OF THE EXISTENCE OF MATERIAL THINGS, AND OF THE REAL DISTINCTION BETWEEN THE MIND AND BODY OF MAN.

There now only remains the inquiry as to whether material things exist. With regard to this question, I at least know with certainty that such things may exist, in as far as they constitute the object of the pure mathematics, since, regarding them in this aspect, I can conceive them clearly and distinctly. For there can be no doubt that God possesses the power of producing all the objects I am able distinctly to conceive, and I never considered anything impossible to him, unless when I experienced a contradiction in the attempt to conceive it aright. Further, the faculty of imagination which I possess, and of which I am conscious that I make use when I apply myself to the consideration of material things, is sufficient to persuade me of their existence: for, when I attentively consider what imagination is, I find that it is simply a certain application of the cognitive faculty (*facultas cognoscitiva*) to a body which is immediately present to it, and which therefore exists.

And to render this quite clear, I remark, in the first place, the difference that subsists between imagination and pure intellection [or conception]. For example, when I imagine a triangle I not only conceive (*intelligo*) that it is a figure comprehended by three lines, but at the same time also I look upon (*intueor*) these three lines as present by the power and internal application of my mind (*acie mentis*), and this is what I call imagining. But if I desire to think of a chiliogon, I indeed rightly conceive that it is a figure composed of a thousand sides, as easily as I conceive that a triangle is a figure composed of only three sides; but I cannot imagine the thousand sides of a chiliogon as I do the three sides of a triangle, nor, so to speak, view them as present [with the eyes of my mind]. And although, in accordance with the habit I have of always imagining something when I think of corporeal things, it may happen that, in conceiving a chiliogon, I confusedly represent some figure to myself, yet it is quite evident that this is not a chiliogon, since it in no wise differs from that which I would represent to myself, if I were to think of a myriogon, or any other figure of many sides; nor would this representation be of any use in discovering and unfolding the properties that constitute the difference between a chiliogon and other polygons. But if the question turns on a pentagon, it is quite true that I can conceive its figure, as well as that of a chiliogon, without the aid of imagination; but I can likewise imagine it by applying the attention of my mind to its five sides, and at the same time to the area which they contain. Thus I observe that a special effort of mind is necessary to the act of imagination, which is not required to conceiving or understanding (*ad intelligendum*); and this special exertion of mind clearly shows the difference between imagination and pure intellection (*imaginatio et intellectio pura*). I remark, besides, that this power of imagination which I possess, in as far as it differs from the power of conceiving, is in no way necessary to my [nature or] essence, that is, to the essence of my mind; for although I did not possess it, I should still remain the same that I now am, from which it seems we may conclude that it depends on something different from the mind. And I easily understand that, if some body exists, with which my mind is so conjoined and united as to be able, as it were, to consider it when it chooses, it may thus imagine corporeal objects; so that this mode of thinking differs from pure intellection only in this respect, that the mind in conceiving turns in some way upon itself, and considers some one of the ideas it possesses within itself; but in imagining it turns towards the body, and contemplates in it some object conformed to the idea which it either of itself conceived or apprehended by sense. I easily understand, I say, that imagination may be thus formed, if it is true that there are bodies; and because I find no other obvious mode of explaining it, I thence, with probability, conjecture that they exist, but only with probability; and although I carefully examine all things, nevertheless I do not find that, from the distinct idea of corporeal nature I have in my imagination, I can necessarily infer the existence of any body.

But I am accustomed to imagine many other objects besides that corporeal nature which is the object of the pure mathematics, as, for example, colours, sounds, tastes, pain, and the like, although with less distinctness; and, inasmuch as I perceive these objects much better by the senses, through the medium of which and of memory, they seem to have reached the imagination, I believe that, in order the more advantageously to examine them, it is proper I should at the same time examine what sense-perception is, and inquire whether from those ideas that are apprehended by this mode of thinking (consciousness), I cannot obtain a certain proof of the existence of corporeal objects.

And, in the first place, I will recall to my mind the things I have hitherto held as true, because perceived by the senses, and the foundations upon which my belief in their truth rested; I will, in the second place, examine the reasons that afterwards constrained me to doubt of them; and, finally, I will consider what of them I ought now to believe.

Firstly, then, I perceived that I had a head, hands, feet, and other members composing that body which I considered as part, or perhaps even as the whole, of myself. I perceived further, that that body was placed among many others, by which it was capable of being affected in diverse ways, both beneficial and hurtful; and what was beneficial I remarked by a certain sensation of pleasure, and what was hurtful by a sensation of pain. And, besides this pleasure and pain, I was likewise conscious of hunger, thirst, and other appetites, as well as certain corporeal inclinations towards joy, sadness, anger, and similar passions. And, out of myself, besides the extension, figure, and motions of bodies, I likewise perceived in them hardness, heat, and the other tactile qualities, and, in addition, light, colours, odours, tastes, and sounds, the variety of which gave me the means of distinguishing the sky, the earth, the sea, and generally all the other bodies, from one another. And certainly, considering the ideas of all these qualities, which were presented to my mind, and which alone I properly and immediately perceived, it was not without reason that I thought I perceived certain objects wholly different from my thought, namely, bodies from which those ideas proceeded; for I was conscious that the ideas were presented to me without my consent being required, so that I could not perceive any object, however desirous I might be, unless it were present to the organ of sense; and it was wholly out of my power not to perceive it when it was thus present. And because the ideas I perceived by the senses were much more lively and clear, and even, in their own way, more distinct than any of those I could of myself frame by meditation, or which I found impressed on my memory, it seemed that they could not have proceeded from myself, and must therefore have been caused in me by some other objects: and as of those objects I had no knowledge beyond what the ideas themselves gave me, nothing was so likely to occur to my mind as the supposition that the objects were similar to the ideas which they caused. And because I recollected also that I had formerly trusted to the senses, rather than to reason, and that the ideas which I myself formed were not so clear as those I perceived by sense, and that they were even for the most part composed of parts of the latter, I was readily persuaded that I had no idea in my intellect which had not formerly passed through the senses. Nor was I altogether wrong in likewise believing that that body which, by a special right, I called my own, pertained to me more properly and strictly than any of the others; for in truth, I could never be separated from it as from other bodies: I felt in it and on account of it all my appetites and affections, and in fine I was affected in its parts by pain and the titillation of pleasure, and not in the parts of the other bodies that were separated from it. But when I inquired into the reason why, from this I know not what sensation of pain, sadness of mind should follow, and why from the sensation of pleasure joy should arise, or why this indescribable twitching of the stomach, which I call hunger, should put me in mind of taking food, and the parchedness of the throat of drink, and so in other cases, I was unable to give any explanation, unless that I was so taught by nature; for there is assuredly no affinity, at least none that I am able to comprehend, between this irritation of the stomach and the desire of food, any more than between the perception of an object that causes pain and the consciousness of sadness which springs from the perception. And in the same way it seemed to me that all the other judgments I had formed regarding the objects of sense, were dictates of nature; because I remarked that those judgments were formed in me, before I had leisure to weigh and consider the reasons that might constrain me to form them.

But, afterwards, a wide experience by degrees sapped the faith I had reposed in my senses; for I frequently observed that towers, which at a distance seemed round, appeared square when more closely viewed, and that colossal figures, raised on the summits of these towers, looked like small statues, when viewed from the bottom of them; and, in other instances without number, I also discovered error in judgments founded on the external senses; and not only in those founded on the external, but even in those that rested on the internal senses; for is there aught more internal than pain? and yet I have sometimes been informed by parties whose arm or leg had been amputated, that they still occasionally seemed to feel pain in that part of the body which they had lost,—a circumstance that led me to think that I could not be quite certain even that any one of my members was affected when I felt pain in it. And to these grounds of doubt I shortly afterwards also added two others of very wide generality: the first of them was that I believed I never perceived anything when awake which I could not occasionally think I also perceived when asleep, and as I do not believe that the ideas I seem to perceive in my sleep proceed from objects external to me, I did not any more observe any ground for believing this of such as I seem to perceive when awake; the second was that since I was as yet ignorant of the author of my being, or at least supposed myself to be so, I saw nothing to prevent my having been so constituted by nature as that I should be deceived even in matters that appeared to me to possess the greatest truth. And, with respect to the grounds on which I had before been persuaded of the existence of sensible objects, I had no great difficulty in finding suitable answers to them; for as nature seemed to incline me to many things from which reason made me averse, I thought that I ought not to confide much in its teachings. And although the perceptions of the senses were not dependent on my will, I did not think that I ought on that ground to conclude that they proceeded from things different from myself, since perhaps there might be found in me some faculty, though hitherto unknown to me, which produced them.

But now that I begin to know myself better, and to discover more clearly the author of my being, I do not, indeed, think that I ought rashly to admit all which the senses seem to teach, nor, on the other hand, is it my conviction that I ought to doubt in general of their teachings.

And, firstly, because I know that all which I clearly and distinctly conceive can be produced by God exactly as I conceive it, it is sufficient that I am able clearly and distinctly to conceive one thing apart from another, in order to be certain that the one is different from the other, seeing they may at least be made to exist separately, by the omnipotence of God; and it matters not by what power this separation is made, in order to be compelled to judge them different; and, therefore, merely because I know with certitude that I exist, and because, in the meantime, I do not observe that aught necessarily belongs to my nature or essence beyond my being a thinking thing, I rightly conclude that my essence consists only in my being a thinking thing, [or a substance whose whole essence or nature is merely thinking]. And although I may, or rather, as I will shortly say, although I certainly do possess a body with which I am very closely conjoined; nevertheless, because, on the one hand, I have a clear and distinct idea of myself, in as far as I am only a thinking and unextended thing, and as, on the other hand, I possess a distinct idea of body, in as far as it is only an extended and unthinking thing, it is certain that I, [that is, my mind, by which I am what I am], is entirely and truly distinct from my body, and may exist without it.

Moreover, I find in myself diverse faculties of thinking that have each their special mode: for example, I find I possess the faculties of imagining and perceiving, without which I can indeed clearly and distinctly conceive myself as entire, but I cannot reciprocally conceive them without conceiving myself, that is to say, without an intelligent substance in which they reside, for [in the notion we have of them, or to use the terms of the schools] in their formal concept, they comprise some sort of intellection; whence I perceive that they are distinct from myself as modes are from things. I remark likewise certain other faculties, as the power of changing place, of assuming diverse figures, and the like, that cannot be conceived and cannot therefore exist, any more than the preceding, apart from a substance in which they inhere. It is very evident, however, that these faculties, if they really exist, must belong to some corporeal or extended substance, since in their clear and distinct concept there is contained some sort of extension, but no intellection at all. Farther, I cannot doubt but that there is in me a certain passive faculty of perception, that is, of receiving and taking knowledge of the ideas of sensible things; but this would be useless to me, if there did not also exist in me, or in some other thing, another active faculty capable of forming and producing those ideas. But this active faculty cannot be in me [in as far as I am but a thinking thing], seeing that it does not presuppose thought, and also that those ideas are frequently produced in my mind without my contributing to it in any way, and even frequently contrary to my will. This faculty must therefore exist in some substance different from me, in which all the objective reality of the ideas that are produced by this faculty, is contained formally or eminently, as I before remarked: and this substance is either a body, that is to say, a corporeal nature in which is contained formally [and in effect] all that is objectively [and by representation] in those ideas; or it is God himself, or some other creature, of a rank superior to body, in which the same is contained eminently. But as God is no deceiver, it is manifest that he does not of himself and immediately communicate those ideas to me, nor even by the intervention of any creature in which their objective reality is not formally, but only eminently, contained. For as he has given me no faculty whereby I can discover this to be the case, but, on the contrary, a very strong inclination to believe that those ideas arise from corporeal objects, I do not see how he could be vindicated from the charge of deceit, if in truth they proceeded from any other source, or were produced by other causes than corporeal things: and accordingly it must be concluded, that corporeal objects exist. Nevertheless they are not perhaps exactly such as we perceive by the senses, for their comprehension by the senses is, in many instances, very obscure and confused; but it is at least necessary to admit that all which I clearly and distinctly conceive as in them, that is, generally speaking, all that is comprehended in the object of speculative geometry, really exists external to me.

But with respect to other things which are either only particular, as, for example, that the sun is of such a size and figure, etc., or are conceived with less clearness and distinctness, as light, sound, pain, and the like, although they are highly dubious and uncertain, nevertheless on the ground alone that God is no deceiver, and that consequently he has permitted no falsity in my opinions which he has not likewise given me a faculty of correcting, I think I may with safety conclude that I possess in myself the means of arriving at the truth. And, in the first place, it cannot be doubted that in each of the dictates of nature there is some truth: for by nature, considered in general, I now understand nothing more than God himself, or the order and disposition established by God in created things; and by my nature in particular I understand the assemblage of all that God has given me.

But there is nothing which that nature teaches me more expressly [or more sensibly] than that I have a body which is ill affected when I feel pain, and stands in need of food and drink when I experience the sensations of hunger and thirst, etc. And therefore I ought not to doubt but that there is some truth in these informations.

Nature likewise teaches me by these sensations of pain, hunger, thirst, etc., that I am not only lodged in my body as a pilot in a vessel, but that I am besides so intimately conjoined, and as it were intermixed with it, that my mind and body compose a certain unity. For if this were not the case, I should not feel pain when my body is hurt, seeing I am merely a thinking thing, but should perceive the wound by the understanding alone, just as a pilot perceives by sight when any part of his vessel is damaged ; and when my body has need of food or drink, I should have a clear knowledge of this, and not be made aware of it by the confused sensations of hunger and thirst: for, in truth, all these sensations of hunger, thirst, pain, etc., are nothing more than certain confused modes of thinking, arising from the union and apparent fusion of mind and body.

Besides this, nature teaches me that my own body is surrounded by many other bodies, some of which I have to seek after, and others to shun. And indeed, as I perceive different sorts of colours, sounds, odours, tastes, heat, hardness, etc., I safely conclude that there are in the bodies from which the diverse perceptions of the senses proceed, certain varieties corresponding to them, although, perhaps, not in reality like them ; and since, among these diverse perceptions of the senses, some are agreeable, and others disagreeable, there can be no doubt that my body, or rather my entire self, in as far as I am composed of body and mind, may be variously affected, both beneficially and hurtfully, by surrounding bodies.

But there are many other beliefs which, though seemingly the teaching of nature, are not in reality so, but which obtained a place in my mind through a habit of judging inconsiderately of things. It may thus easily happen that such judgments shall contain error: thus, for example, the opinion I have that all space in which there is nothing to affect [or make an impression on] my senses is void; that in a hot body there is something in every respect similar to the idea of heat in my mind; that in a white or green body there is the same whiteness or greenness which I perceive; that in a bitter or sweet body there is the same taste, and so in other instances; that the stars, towers, and all distant bodies, are of the same size and figure as they appear to our eyes, etc. But that I may avoid everything like indistinctness of conception, I must accurately define what I properly understand by being taught by nature. For nature is here taken in a narrower sense than when it signifies the sum of all the things which God has given me; seeing that in that meaning the notion comprehends much that belongs only to the mind [to which I am not here to be understood as referring when I use the term nature]; as, for example, the notion I have of the truth, that what is done cannot be undone, and all the other truths I discern by the natural light [without the aid of the body]; and seeing that it comprehends likewise much besides that belongs only to body, and is not here any more contained under the name nature, as the quality of heaviness, and the like, of which I do not speak,—the term being reserved exclusively to designate the things which God has given to me as a being composed of mind and body. But nature, taking the term in the sense explained, teaches me to shun what causes in me the sensation of pain, and to pursue what affords me the sensation of pleasure, and other things of this sort; but I do not discover that it teaches me, in addition to this, from these diverse perceptions of the senses, to draw any conclusions respecting external objects without a previous [careful and mature] consideration of them by the mind: for it is, as appears to me, the office of the mind alone, and not of the composite whole of mind and body, to discern the truth in those matters. Thus, although the impression a star makes on my eye is not larger than that from the flame of a candle, I do not, nevertheless, experience any real or positive impulse determining me to believe that the star is not greater than the flame; the true account of the matter being merely that I have so judged from my youth without any rational ground. And, though on approaching the fire I feel heat, and even pain on approaching it too closely, I have, however, from this no ground for holding that something resembling the heat I feel is in the fire, any more than that there is something similar to the pain; all that I have ground for believing is, that there is something in it, whatever it may be, which excites in me those sensations of heat or pain. So also, although there are spaces in which I find nothing to excite and affect my senses, I must not therefore conclude that those spaces contain in them no body; for I see that in this, as in many other similar matters, I have been accustomed to pervert the order of nature, because these perceptions of the senses, although given me by nature merely to signify to my mind what things are beneficial and hurtful to the composite whole of which it is a part, and being sufficiently clear and distinct for that purpose, are nevertheless used by me as infallible rules by which to determine immediately the essence of the bodies that exist out of me, of which they can of course afford me only the most obscure and confused knowledge.

But I have already sufficiently considered how it happens that, notwithstanding the supreme goodness of God, there is falsity in my judgments. A difficulty, however, here presents itself, respecting the things which I am taught by nature must be pursued or avoided, and also respecting the internal sensations in which I seem to have occasionally detected error, [and thus to be directly deceived by nature]: thus, for example, I may be so deceived by the agreeable taste of some viand with which poison has been mixed, as to be induced to take the poison. In this case, however, nature may be excused, for it simply leads me to desire the viand for its agreeable taste, and not the poison, which is unknown to it; and thus we can infer nothing from this circumstance beyond that our nature is not omniscient; at which there is assuredly no ground for surprise, since, man being of a finite nature, his knowledge must likewise be of limited perfection. But we also not unfrequently err in that to which we are directly impelled by nature, as is the case with invalids who desire drink or food that would be hurtful to them. It will here, perhaps, be alleged that the reason why such persons are deceived is that their nature is corrupted; but this leaves the difficulty untouched, for a sick man is not less really the creature of God than a man who is in full health; and therefore it is as repugnant to the goodness of God that the nature of the former should be deceitful as it is for that of the latter to be so. And, as a clock, composed of wheels and counter weights, observes not the less accurately all the laws of nature when it is ill made, and points out the hours incorrectly, than when it satisfies the desire of the maker in every respect; so likewise if the body of man be considered as a kind of machine, so made up and composed of bones, nerves, muscles, veins, blood, and skin, that although there were in it no mind, it would still exhibit the same motions which it at present manifests involuntarily, and therefore without the aid of the mind, [and simply by the dispositions of its organs], I easily discern that it would also be as natural for such a body, supposing it dropsical, for example, to experience the parchedness of the throat that is usually accompanied in the mind by the sensation of thirst, and to be disposed by this parchedness to move its nerves and its other parts in the way required for drinking, and thus increase its malady and do itself harm, as it is natural for it, when it is not indisposed to be stimulated to drink for its good by a similar cause; and although looking to the use for which a clock was destined by its maker, I may say that it is deflected from its proper nature when it incorrectly indicates the hours, and on the same principle, considering the machine of the human body as having been formed by God for the sake of the motions which it usually manifests, although I may likewise have ground for thinking that it does not follow the order of its nature when the throat is parched and drink does not tend to its preservation, nevertheless I yet plainly discern that this latter acceptation of the term nature is very different from the other; for this is nothing more than a certain denomination, depending entirely on my thought, and hence called extrinsic, by which I compare a sick man and an imperfectly constructed clock with the idea I have of a man in good health and a well made clock; while by the other acceptation of nature is understood something which is truly found in things, and therefore possessed of some truth.

But certainly, although in respect of a dropsical body, it is only by way of exterior denomination that we say its nature is corrupted, when, without requiring drink, the throat is parched; yet, in respect of the composite whole, that is, of the mind in its union with the body, it is not a pure denomination, but really an error of nature, for it to feel thirst when drink would be hurtful to it: and, accordingly, it still remains to be considered why it is that the goodness of God does not prevent the nature of man thus taken from being fallacious.

To commence this examination accordingly, I here remark, in the first place, that there is a vast difference between mind and body, in respect that body, from its nature, is always divisible, and that mind is entirely indivisible. For in truth, when I consider the mind, that is, when I consider myself in so far only as I am a thinking thing, I can distinguish in myself no parts, but I very clearly discern that I am somewhat absolutely one and entire; and although the whole mind seems to be united to the whole body, yet, when a foot, an arm, or any other part is cut off, I am conscious that nothing has been taken from my mind ; nor can the faculties of willing, perceiving, conceiving, etc., properly be called its parts, for it is the same mind that is exercised [all entire] in willing, in perceiving, and in conceiving, etc. But quite the opposite holds in corporeal or extended things; for I cannot imagine any one of them [how small soever it may be], which I cannot easily sunder in thought, and which, therefore, I do not know to be divisible. This would be sufficient to teach me that the mind or soul of man is entirely different from the body, if I had not already been apprised of it on other grounds.

I remark, in the next place, that the mind does not immediately receive the impression from all the parts of the body, but only from the brain, or perhaps even from one small part of it, viz., that in which the common sense (*sensus communis*) is said to be, which as often as it is affected in the same way, gives rise to the same perception in the mind, although meanwhile the other parts of the body may be diversely disposed, as is proved by innumerable experiments, which it is unnecessary here to enumerate.

I remark, besides, that the nature of body is such that none of its parts can be moved by another part a little removed from the other, which cannot likewise be moved in the same way by any one of the parts that lie between those two, although the most remote part does not act at all. As, for example, in the cord A, B, C, D, [which is in tension], if its last part D, be pulled, the first part A, will not be moved in a different way than it would be were one of the intermediate parts B or C to be pulled, and the last part D meanwhile to remain fixed. And in the same way, when I feel pain in the foot, the science of physics teaches me that this sensation is experienced by means of the nerves dispersed over the foot, which, extending like cords from it to the brain, when they are contracted in the foot, contract at the same time the inmost parts of the brain in which they have their origin, and excite in these parts a certain motion appointed by nature to cause in the mind a sensation of pain, as if existing in the foot: but as these nerves must pass through the tibia, the leg, the loins, the back, and neck, in order to reach the brain, it may happen that although their extremities in the foot are not affected, but only certain of their parts that pass through the loins or neck, the same movements, nevertheless, are excited in the brain by this motion as would have been caused there by a hurt received in the foot, and hence the mind will necessarily feel pain in the foot, just as if it had been hurt; and the same is true of all the other perceptions of our senses.

I remark, finally, that as each of the movements that are made in the part of the brain by which the mind is immediately affected, impresses it with but a single sensation, the most likely supposition in the circumstances is, that this movement causes the mind to experience, among all the sensations which it is capable of impressing upon it, that one which is the best fitted, and generally the most useful for the preservation of the human body when it is in full health. But experience shows us that all the perceptions which nature has given us are of such a kind as I have mentioned; and accordingly, there is nothing found in them that does not manifest the power and goodness of God. Thus, for example, when the nerves of the foot are violently or more than usually shaken, the motion passing through the medulla of the spine to the innermost parts of the brain affords a sign to the mind on which it experiences a sensation, viz., of pain, as if it were in the foot, by which the mind is admonished and excited to do its utmost to remove the cause of it as dangerous and hurtful to the foot. It is true that God could have so constituted the nature of man as that the same motion in the brain would have informed the mind of something altogether different: the motion might, for example, have been the occasion on which the mind became conscious of itself, in so far as it is in the brain, or in so far as it is in some place intermediate between the foot and the brain, or, finally, the occasion on which it perceived some other object quite different, whatever that might be; but nothing of all this would have so well contributed to the preservation of the body as that which the mind actually feels. In the same way, when we stand in need of drink, there arises from this want a certain parchedness in the throat that moves its nerves, and by means of them the internal parts of the brain; and this movement affects the mind with the sensation of thirst, because there is nothing on that occasion which is more useful for us than to be made aware that we have need of drink for the preservation of our health; and so in other instances.

Whence it is quite manifest that, notwithstanding the sovereign goodness of God, the nature of man, in so far as it is composed of mind and body, cannot but be sometimes fallacious. For, if there is any cause which excites, not in the foot, but in some one of the parts of the nerves that stretch from the foot to the brain, or even in the brain itself, the same movement that is ordinarily created when the foot is ill affected, pain will be felt, as it were, in the foot, and the sense will thus be naturally deceived; for as the same movement in the brain can but impress the mind with the same sensation, and as this sensation is much more frequently excited by a cause which hurts the foot than by one acting in a different quarter, it is reasonable that it should lead the mind to feel pain in the foot rather than in any other part of the body. And if it sometimes happens that the parchedness of the throat does not arise, as is usual, from drink being necessary for the health of the body, but from quite the opposite cause, as is the case with the dropsical, yet it is much better that it should be deceitful in that instance, than if, on the contrary, it were continually fallacious when the body is well-disposed; and the same holds true in other cases.

And certainly this consideration is of great service, not only in enabling me to recognise the errors to which my nature is liable, but likewise in rendering it more easy to avoid or correct them: for, knowing that all my senses more usually indicate to me what is true than what is false, in matters relating to the advantage of the body, and being able almost always to make use of more than a single sense in examining the same object, and besides this, being able to use my memory in connecting present with past knowledge, and my understanding which has already discovered all the causes of my errors, I ought no longer to fear that falsity may be met with in what is daily presented to me by the senses. And I ought to reject all the doubts of those bygone days, as hyperbolical and ridiculous, especially the general uncertainty respecting sleep, which I could not distinguish from the waking state: for I now find a very marked difference between the two states, in respect that our memory can never connect our dreams with each other and with the course of life, in the way it is in the habit of doing with events that occur when we are awake. And, in truth, if some one, when I am awake, appeared to me all of a sudden and as suddenly disappeared, as do the images I see in sleep, so that I could not observe either whence he came or whither he went, I should not without reason esteem it either a spectre or phantom formed in my brain, rather than a real man. But when I perceive objects with regard to which I can distinctly determine both the place whence they come, and that in which they are, and the time at which they appear to me, and when, without interruption, I can connect the perception I have of them with the whole of the other parts of my life, I am perfectly sure that what I thus perceive occurs while I am awake and not during sleep. And I ought not in the least degree to doubt of the truth of those presentations, if, after having called together all my senses, my memory, and my understanding for the purpose of examining them, no deliverance is given by any one of these faculties which is repugnant to that of any other: for since God is no deceiver, it necessarily follows that I am not herein deceived. But because the necessities of action frequently oblige us to come to a determination before we have had leisure for so careful an examination, it must be confessed that the life of man is frequently obnoxious to error with respect to individual objects; and we must, in conclusion, acknowledge the weakness of our nature.

Descartes Time Line

1596 31 March: born at La Haye near Tours

1606-14 He attends Jesuit college of La Fleche in Anjou

1616 Descartes takes doctor of law at University of Poitiers

1618 He goes to Holland and joins the army of Prince Maurice of Nassau

1619 He travels in Germany; on 10 November in Ulm has dream of a unified scientific system based on mathematics

1622 He returns to France, during next few years spends time in Paris, but also travels in Europe

1628 He composes *Rules for the Direction of Mind* (which was first published in 1701);

1628 In November Descartes distinguished himself in a confrontation with Chandoux, who claimed that all science is based on probability while Descartes claimed that only certainty could be the basis of human knowledge and that he had a method for attaining certainty. Following this, Descartes was encouraged by Cardinal Berulle to develop his system.

1628 Descartes leaves for Holland which is to be his home until 1649

1629 He begins working on *The World*

1633 The condemnation of Galileo leads Descartes to abandon plans to publish *The World*

1635 The birth of Descartes' natural daughter, named Francine, baptized 7 August (died 1640)

1637 Descartes publishes *Discourse on Method*, with *Optics*, *Meterology* and the *Geometry*

1641 *Meditations on First Philosophy* published together with the first six sets of *Objections and Replies*

1642 Second edition of the *Meditations* published along with all seven sets of objections and replies. Descartes meets Princess Elisabeth of Bohemia

1643 Cartesian philosophy condemned at the University of Utrecht; Descartes' long correspondence with the Princess Elisabeth of Bohemia begins.

1644 Visits France: *The Principles of Philosophy* published.

1647 Descartes is awarded a pension by the King of France; he publishes *Comments on a Certain Broadsheet*; begins work on *The Description of the Human Body*

1648 He is interviewed by Frans Burman at Egmond-Binnen which leads to the publication of *Conversations with Burman*

1649 He goes to Sweden on invitation of Queen Christina; the *Passions of the Soul* published

1650 11 February: dies in Stockholm of pneumonia